Ground Water Recharge Using Waters of Impaired Quality

Committee on Ground Water Recharge

Water Science and Technology Board

Commission on Geosciences, Environment, and Resources

National Academy Press
Washington, D. C. 1994

NOTICE: The project that is the subject of this report was approved by the Governing Board of the National Research Council, whose members are drawn from the councils of the National Academy of Sciences, the National Academy of Engineering, and the Institute of Medicine. The members of the committee responsible for the report were chosen for their special competencies and with regard for appropriate balance.

This report has been reviewed by a group other than the authors according to procedures approved by a Report Review Committee consisting of members of the National Academy of Sciences, the National Academy of Engineering, and the Institute of Medicine.

Support for this project was provided by the Bureau of Reclamation Grant No. 1-FG-81-18250, U.S. Environmental Protection Agency Grant No. CX-818588-01-0, West Basin Municipal Water District, Water Replenishment District of Southern California, Orange County Water District, and National Water Research Institute.

Library of Congress Catalog Card No. 94-66774
International Standard Book Number 0-309-05142-8

Additional copies of this report are available from:

National Academy Press
2101 Constitution Avenue, N.W.
Box 285
Washington, D.C. 20055
800-624-6242, 202-334-3313 (in the Washington Metropolitan Area)

B-463

Copyright 1994 by the National Academy of Sciences. All rights reserved.

Original cover art by Marilyn Kirkman, Arati Artists Gallery, Colorado Springs, Colorado

Printed in the United States of America

TD
404
.G76
1994

COMMITTEE ON GROUND WATER RECHARGE

JULIAN ANDELMAN, *Chair*, University of Pittsburgh
HERMAN BOUWER, U.S. Water Conservation Laboratory, Phoenix, Arizona
RANDALL CHARBENEAU, University of Texas at Austin
RUSSELL CHRISTMAN, University of North Carolina, Chapel Hill
JAMES CROOK, Black & Veatch, Cambridge, Massachusetts
ANNA FAN, California Environmental Protection Agency, Berkeley
DENISE FORT, University of New Mexico
WILFORD GARDNER, University of California, Berkeley
WILLIAM JURY, University of California, Riverside
DAVID MILLER, Geraghty & Miller, Inc., Plainview, New York
ROBERT PITT, University of Alabama at Birmingham
GORDON ROBECK, Water Consultant, Laguna Hills, California
    (until February 1993, see page vii)
HENRY VAUX, JR., University of California, Berkeley
JOHN VECCHIOLI, U.S. Geological Survey, Tallahassee, Florida
MARYLYNN YATES, University of California, Riverside

National Research Council Staff

CHRIS ELFRING, Study Director
ANITA HALL, Project Assistant
ETAN GUMERMAN, Research Intern
ROSEANNE PRICE, Editor

WATER SCIENCE AND TECHNOLOGY BOARD

DANIEL A. OKUN, *Chair*, University of North Carolina, Chapel Hill
A. DAN TARLOCK, *Vice Chair*, Illinois Institute of Technology, Chicago-Kent College of Law
J. DAVID ALLEN, Chevron U.S.A., Inc., New Orleans, Loiusiana
PATRICK L. BREZONIK, University of Minnesota, St. Paul
KENNETH D. FREDERICK, Resources for the Future, Washington, D.C.
DAVID L. FREYBERG, Stanford University, Stanford, California
WILFORD R. GARDNER, University of California, Berkeley
LYNN R. GOLDMAN, California Department of Health Services, Emeryville, California
WILLIAM L. GRAF, Arizona State University, Tempe
THOMAS M. HELLMAN, Bristol-Myers Squibb Company, New York, New York
ROBERT J. HUGGETT, College of William and Mary, Gloucester Point, Virginia
CHARLES C. JOHNSON, Jr., U.S. Public Health Service, Washington, D.C. (Retired)
WILLIAM M. LEWIS, JR., University of Colorado, Boulder
CAROLYN H. OLSEN, Brown and Caldwell, Atlanta, Georgia
CHARLES R. O'MELIA, Johns Hopkins University, Baltimore, Maryland
STAVROS S. PAPADOPULOS, S. S. Papadopulos & Associates, Inc., Rockville, Maryland
BRUCE E. RITTMANN, Northwestern University, Evanston, Illinois
JOY B. ZEDLER, San Diego State University, San Diego

Staff

STEPHEN D. PARKER, Director
SHEILA D. DAVID, Senior Staff Officer
CHRIS ELFRING, Senior Staff Officer
GARY KRAUSS, Staff Officer
JACQUELINE MACDONALD, Staff Officer
JEANNE AQUILINO, Administrative Specialist
ANITA A. HALL, Administrative Assistant
GREGORY NYCE, Senior Project Assistant
MARY BETH MORRIS, Senior Project Assistant
ANGELA BRUBAKER, Project Assistant
ETAN GUMERMAN, Research Intern

COMMISSION ON GEOSCIENCES, ENVIRONMENT, AND RESOURCES

M. GORDON WOLMAN, *Chair*, The Johns Hopkins University, Baltimore, Maryland
PATRICK R. ATKINS, Aluminum Company of America, Pittsburgh, Pennsylvania
PETER EAGLESON, Massachusetts Institute of Technology, Cambridge
EDWARD A. FRIEMAN, Scripps Institution of Oceanography, La Jolla, California
W. BARCLAY KAMB, California Institute of Technology, Pasadena
JACK E. OLIVER, Cornell University, Ithaca, New York
FRANK L. PARKER, Vanderbilt University, Nashville, Tennessee
RAYMOND A. PRICE, Queen's University at Kingston, Ontario
THOMAS A. SCHELLING, University of Maryland, College Park, Maryland
LARRY L. SMARR, University of Illinois, Urbana-Champaign
STEVEN M. STANLEY, The Johns Hopkins University, Baltimore, Maryland
VICTORIA J. TSCHINKEL, Landers and Parsons, Tallahassee, Florida
WARREN WASHINGTON, National Center for Atmospheric Research, Boulder, Colorado
EDITH BROWN WEISS, Georgetown University Law Center, Washington, D.C.

Staff

STEPHEN RATTIEN, Executive Director
STEPHEN D. PARKER, Associate Executive Director
MORGAN GOPNIK, Assistant Executive Director
JEANETTE SPOON, Administrative Officer
SANDI FITZPATRICK, Administrative Associate
ROBIN ALLEN, Senior Project Assistant

The National Academy of Sciences is a private, nonprofit, self-perpetuating society of distinguished scholars engaged in scientific and engineering research, dedicated to the furtherance of science and technology and to their use for the general welfare. Upon the authority of the charter granted to it by the Congress in 1863, the Academy has a mandate that requires it to advise the federal government on scientific and technical matters. Dr. Bruce Alberts is president of the National Academy of Sciences.

The National Academy of Engineering was established in 1964, under the charter of the National Academy of Sciences, as a parallel organization of outstanding engineers. It is autonomous in its administration and in the selection of its members, sharing with the National Academy of Sciences the responsibility for advising the federal government. The National Academy of Engineering also sponsors engineering programs aimed at meeting national needs, encourages education and research, and recognizes the superior achievements of engineers. Dr. Robert M. White is president of the National Academy of Engineering.

The Institute of Medicine was established in 1970 by the National Academy of Sciences to secure the services of eminent members of appropriate professions in the examination of policy matters pertaining to the health of the public. The Institute acts under the responsibility given to the National Academy of Sciences by its congressional charter to be an adviser to the federal government and, upon its own initiative, to identify issues of medical care, research, and education. Dr. Kenneth I. Shine is president of the Institute of Medicine.

The National Research Council was organized by the National Academy of Sciences in 1916 to associate the broad community of science and technology with the Academy's purposes of furthering knowledge and advising the federal government. Functioning in accordance with general policies determined by the Academy, the Council has become the principal operating agency of both the National Academy of Sciences and the National Academy of Engineering in providing services to the government, the public, and the scientific and engineering communities. The Council is administered jointly by both Academies and the Institute of Medicine. Dr. Bruce Alberts and Dr. Robert M. White are chairman and vice chairman, respectively, of the National Research Council.

Dedication

This volume is dedicated to Gordon Robeck, a member of the Committee on Ground Water Recharge, a former member of the Water Science and Technology Board, and a long-time leader in the water community, who passed away in February 1993. A member of the National Academy of Engineering and an internationally respected scientist and engineer, Mr. Robeck had a long and truly distinguished career. He was a pioneer in the provision of safe drinking water. Mr. Robeck will be greatly missed, but his contributions continue to benefit all.

Mr. Robeck spent most of his life in public service, first as a researcher with the U.S. Public Health Service and later as director of the Environmental Protection Agency's Drinking Water Research Division. During his exceptional career, he received many awards, including the Meritorious Service Award from the Public Health Service in 1971; the American Water Works Association Medal for Outstanding Service in 1979; and the EPA Gold Medal for Exceptional Service in 1978. As noted by his friend James M. Symons:

"Gordon received many professional honors, and he enjoyed them, but he never sought them, nor thought them too important. What he did think was important was to have an impact — to make a difference... And he did make a difference: for me, for Cincinnati, for the field, and for all those people who now have better quality drinking water thanks to his efforts."

Preface

Water is increasingly in short supply, especially in the arid and semiarid West, and we as a nation are in a continual search for innovative ways to improve the efficiency with which we manage this critical resource. The growing demand for water, both in the United States and elsewhere around the world, has brought an increasing appreciation for the earth's vast ground water supplies. We look to underground aquifers not just as sources of supply, but as vast storage facilities that give us great management flexibility at relatively affordable cost. One element of a strategy to improve our management of ground water resources is the use of artificial recharge—where excess water is directed purposely into the ground to rebuild or augment ground water supplies. As artificial recharge has increased in popularity, managers have begun to search for additional sources of recharge water. A critical question is whether we might be able to use waters of impaired quality—given appropriate pretreatment, posttreatment, and treatment gained from soil and aquifer processes—to expand our capability to carry out artificial recharge and whether the water recovered from such systems is suitable for potable as well as nonpotable uses.

The Committee on Ground Water Recharge was established by the Water Science and Technology Board of the National Research Council to study the potential of artificial recharge of ground water using source waters of impaired quality, specifically treated municipal wastewater, stormwater runoff, and irrigation return flow. The issues addressed include source water characteristics, treatment technologies, health effects, fate and transport of contaminants, and the sustainability of recharge systems. This report is our attempt to compile a general guide that might be of assistance to federal, state, and local officials and water managers as they face decisions about the feasibility of proposed recharge projects.

The Committee on Ground Water Recharge consisted of 15 members with experience in engineering, soil science, hydrology, public health, microbiology, economics, law, and other related fields. We gained insights from a far larger group by inviting guests to our meetings, conducting case studies, and reviewing the literature at great length. Appendix A acknowledges some of the community who assisted our project. In particular, however, I want to express my great appreciation to each committee member—each gave significant time and energy to create this report. I also want to thank the staff of the Water Science and Technology Board, especially Chris Elfring, study director, and Anita Hall, project assistant. I would also like to thank the study's sponsors: the U.S. Environmental Protection Agency, the U.S. Bureau of Reclamation, the Water Replenishment District of Southern California, California's Orange County Water District, California's West Basin Municipal Water District, and the National Water Research Institute. Without this support, the study could not have been completed.

The committee's deliberations touch on many issues. The recommendations focus on broad issues, rather than the site-specific details associated with the great variety of possible recharge locations, source waters, and regions. We hope that our report will help move the nation forward in its ability to benefit from the potential offered by artificial recharge.

> Julian Andelman, Chair
> Committee on Ground Water Recharge

Contents

SUMMARY   1

1   AN INTRODUCTION TO ARTIFICIAL RECHARGE   12
    A Primer on Artificial Recharge, 13
    Environmental Effects, 30
    Summary, 33
    References, 34

2   SOURCE WATERS AND THEIR TREATMENT   35
    Municipal Wastewater, 36
    Disinfection, 52
    Urban Stormwater Runoff, 60
    Irrigation Return Flow, 78
    Summary, 86
    References, 91

3   SOIL AND AQUIFER PROCESSES   97
    Conditions Influencing Pretreatment, 98
    General Description of Subsurface Processes, 100
    Important Soil-Aquifer Properties, 109
    Undesirable Soil Characteristics, 111
    Transport and Fate of Specific Constituents of Recharge Water, 114
    Sustainability of the SAT System, 122
    Performance and Compliance Monitoring, 124

Summary, 127
References, 128

4 PUBLIC HEALTH ISSUES 132
   Risk Assessment Methodology, Approaches, and
      Interpretation, 134
   Studies of Health Impacts, 136
   Chemical Constituents of Concern, 143
   Microorganisms of Concern, 153
   Risks from Disinfectants and Disinfection By-Products Versus
      Risks from Pathogens, 168
   Health Implications from Nonpotable Uses, 170
   Summary, 173
   References, 175

5 ECONOMIC, LEGAL, AND INSTITUTIONAL
   CONSIDERATIONS 179
   Economic Issues, 179
   Legal Issues, 186
   Institutional Issues, 201
   Public Attitudes Toward the Use of Reclaimed Water, 204
   Summary, 207
   References, 208

6 SELECTED ARTIFICIAL RECHARGE PROJECTS 211
   Water Factory 21, Orange County, California, 212
   Montebello Forebay Ground Water Recharge Project,
      Los Angeles, California, 217
   Phoenix, Arizona, Projects, 222
   El Paso, Texas, Recharge Project, 233
   Long Island, New York, Recharge Basins, 240
   Orlando Area, Florida, Stormwater Drainage Wells, 245
   Dan Region Reclamation and Reuse Project, Metropolitan
      Tel Aviv, Israel, 251
   References, 258

7 CONCLUSIONS AND RECOMMENDATIONS 260
   Artificial Recharge: A Viable Option, 261
   Potential Impaired Quality Sources, 262
   Human Health Concerns, 264
   System Management and Monitoring, 266
   Economic Considerations, 267
   Legal and Institutional Considerations, 268

APPENDIXES

A	Acknowledgements	273
B	Biographical Sketches of Committee Members	275
C	Glossary	279

Ground Water Recharge
Using Waters of
Impaired Quality

Summary

Ground water is a far more important resource than is often realized—excluding the water locked in glaciers and icecaps, about 97 percent of the world's fresh water is ground water, while streams, rivers, and lakes hold only about 3 percent (Bouwer, 1978). In the United States, about one-half of the population and three-fourths of the public water supply systems rely on ground water. Ground water also provides a critical source of water for agricultural irrigation and industries. But as is often the case with critical resources, ground water is not always available when and where needed, especially in water-short areas where heavy use has depleted underground reserves.

The growing competition for water in the United States and elsewhere around the world are leading to even greater use of this enormous water resource. As part of this trend, there has been increased interest in the use of artificial recharge to augment ground water supplies, especially in the western United States and other water-short areas such as Florida. Stated simply, artificial recharge is a process by which excess surface water is directed into the ground—either by spreading on the surface, by using recharge wells, or by altering natural conditions to increase infiltration—to replenish an aquifer. The most common purpose of artificial recharge (also called planned recharge) is to store water underground in times of surplus to meet water demand in times of shortage. Recovered recharge water is especially well-suited to nonpotable uses such as landscape irrigation (which, in turn, frees other supplies for higher uses); under appropriate conditions and where better sources are not available, recovered recharge water also can be an option for potable use. Artificial recharge can be used to control sea water intrusion in coastal aquifers, control land subsidence caused by declining ground water levels, maintain base flow in some streams, and raise ground water levels to reduce the cost of ground water pumping.

As the benefits of artificial recharge of ground water have become evident, water planners have sought alternative sources of water for recharge projects. The National Research Council's Committee on Ground Water Recharge was established to study issues associated with the recharge of ground water using source waters of impaired quality, specifically treated municipal wastewater, stormwater runoff, and irrigation return flow, and issues associated with the use of recovered recharge water for potable as well as nonpotable purposes. (This report does not address industrial wastewater, which can contain too wide and too different an array of possible constituents to be dealt with in this same volume.) The committee was asked to address a range of topics such as source water characteristics, pretreatment and recharge technologies, public health, and the nature of the physical, chemical, and biological processes and transformations that occur during transport through the subsurface. Economic, institutional, and regulatory issues were also to be considered.

The committee is aware that much work has been done in the past to understand the opportunities and potential problems related to artificial recharge of ground water, and we have been careful to try to build on this strong foundation. The committee also recognizes that artificial recharge using waters of impaired quality is one of many strategies that can be used, alone or in conjunction with other strategies, to augment water supplies, such as reducing water consumption or creating secondary water systems that deliver certain wastewaters directly to nonpotable uses (e.g., the use of gray water for landscape irrigation). This report summarizes the state of our knowledge about artificial recharge using source waters of impaired quality and its usefulness and makes recommendations to help the nation use this water management tool more effectively.

SOURCE WATERS AND THEIR TREATMENT

The quality of the source waters used to recharge ground water has a direct bearing on operational aspects of the recharge facilities and also on the ultimate use to be made of the recovered water. In general, the source water characteristics that affect the operational aspects of recharge facilities include suspended solids, dissolved gases, nutrients, biochemical oxygen demand, microorganisms, and the sodium adsorption ratio (which affects soil permeability). The constituents that have the greatest potential effects when potable reuse is being considered include organic and inorganic toxicants, nitrogen compounds, and pathogens.

Of the three types of impaired quality water considered in this report—treated municipal wastewater, stormwater runoff, and irrigation return flow—municipal wastewater is by far the most consistent spatially and temporally and in terms of both quantity and quality. Exceptions to this generalization are where raw municipal wastewater and stormwater are commingled or where there are variable industrial contributions. The major constituents of municipal wastewater have been studied extensively, but less is known about trace constituents. Possible constituents of concern in municipal wastewater effluent include organic compounds, nitrogen species, phosphorus, pathogenic organisms, and suspended solids. Treatment processes are available to bring municipal wastewater effluent to levels acceptable for various recharge applications; however, even when it has been treated to a high degree, effluent that has been disinfected with chlorine will contain disinfection by-products that can be of concern if the recovered water is to be used for potable purposes.

The quality of stormwater runoff is affected by several factors, including rainfall quantity and intensity, the natural and anthropogenic characteristics of the drainage basin, time since the last runoff event, and, in northern areas, time of year. Constituents of concern in stormwater runoff include trace metals, organic compounds, pathogenic organisms, suspended solids, and in northern climates in the winter, dissolved solids such as sodium chloride contributed by road deicing practices. Generally, stormwater runoff from residential areas is of good quality but its quantity may be extremely erratic.

Irrigation return flow exhibits the widest variation in quality of the three potential source waters considered here, and quality characteristics beyond salinity and concentrations of nitrate are not well studied. Salt content can be a problem in arid and semiarid areas, and suspended solids, nutrients, pesticide residues, and trace element concentrations including selenium, uranium, boron, and arsenic may also be of concern. Although treatment processes are available to remove the constituents of concern to acceptable levels, treatment of irrigation return flow generally is not done, and the cost-effectiveness of such treatment is questionable.

SOIL AND AQUIFER PROCESSES

The common assumption that passage of source water through the soil to the aquifer and through the aquifer to the point of withdrawal provides no treatment is overly conservative when applied to most chemicals and microorganisms. The soil and underlying aquifer have a great capacity to remove contaminants and pathogens from recharge water. The ideal soil for a soil-aquifer-treatment (SAT) system balances the need for a high recharge rate, which occurs in coarse-textured soils, with the need for efficient contaminant adsorption and removal, which are better in fine-textured soils.

The unsaturated soil layer (vadose zone) can play an important role in artificial recharge: it can remove or reduce the chemical and biological constituents present in the impaired quality source water as it moves toward the underlying aquifer and thus help reduce potential health risks before the recharge water enters the ground water. Nitrogen, for example, quickly transforms to nitrate, which is very mobile under normal conditions in the soil but which can be removed only by denitrification under anaerobic conditions. Phosphorus levels are reduced by sorption and precipitation, although not completely. Trace metals, with the exception of boron, are strongly attenuated and precipitated in the soil, especially under aerobic and alkaline conditions. Organic chemicals are removed to varying degrees by volatilization or chemical or biological degradation during passage through the vadose zone. Some pathogen removal by filtration occurs for larger organisms, and there is some sorption of bacteria and viruses. Unfortunately, the processes by which removal occurs are not completely efficient in a natural setting, and not all constituents are retained or degraded to the same extent. Moreover, management strategies that may enhance hydraulic capacity or removal of one chemical or pathogen may actually decrease the efficiency of removal of another.

With adequate management and monitoring, an SAT system may reduce pretreatment and posttreatment costs. With proper management, an SAT operation employing periodic drying to reduce clogging should be sustainable indefinitely. Slow trace element migration remains a concern, however; the presence of such constituents necessitates careful monitoring during SAT use and regulation during closure of a facility. Although near-surface monitoring is desirable for proper vigilance, soil variability makes it very difficult to provide complete coverage with existing devices. A combination of near-surface and distant monitoring thus provides the best compromise.

PUBLIC HEALTH ISSUES

A major consideration in the use of impaired-quality waters for artificial recharge is the possible presence of chemical and microbiological agents in the source waters that may be hazardous to human health. Such concerns apply both to potable use and to indirect human exposures that might occur from nonpotable uses, although the possible exposure and thus the risk is significantly less for nonpotable reuse. While a vast body of knowledge exists about relatively uncontaminated, conventional source waters, there is still some uncertainty about the risks associated with impaired sources, principally related to the presence of synthetic organic chemicals, disinfection by-products, and some pathogenic organisms.

The challenge in considering the health risks from recharge systems is to assess and understand the relative risks and develop strategies for the use and operation of recharge systems to minimize those risks. A principal issue is the extent to which impaired-quality source waters require treatment prior to recharge. Another key issue, which is equally important with conventional water supplies, is the need to minimize not only the potential exposures to pathogenic microorganisms, but also the concentrations of possible disinfection by-products. The behavior and fate of microorganisms, disinfection by-products, and other chemical toxicants in the ground water system will affect their concentrations at the point of recovery. Therefore, an understanding of the chemical and microbiological composition of the source water and the changes it can undergo in the complex underground environment is key to the optimal use of impaired-quality water for recharge.

For conventional and impaired-quality sources alike, disinfection by-products and their precursors are of potential concern when the water is intended to support potable uses. The nature and toxicity of disinfection by-products have been studied most widely for chlorine disinfection. The by-products of other disinfectants, such as ozone and chloramine, are not as well characterized. Also, the nature of the disinfection by-products and their precursors in disinfected wastewater is uncertain, as is their behavior in and their effects on ground water aquifers. Recent studies have indicated that disinfection by-products like trihalomethanes and haloacetic acids can be removed by biological degradation in aquifers. Reduction of disinfection by-product precursors during pretreatment will help reduce associated risks. In balancing the risks in using chemical disinfectants to reduce pathogenic microorganisms with those associated with the disinfection by-products formed in the process, it should be emphasized that effective disinfection is critical. The probability of mortality induced by improperly disinfected drinking water exceeds the carcinogenic risks posed by disinfection by-products associated with chlorine by as much as 1,000-fold (Bull et al., 1990).

The public health implications of nonpotable reuse, except where intended for market crops to be eaten raw, have not been addressed as extensively as the implications of potable reuse because nonpotable reuse has been practiced widely for decades without public concern and because the exposure from nonpotable reuse, and thus the risk, is limited.

The health implications of using reclaimed water for potable purposes have been investigated at a number of projects: for example, studies of direct potable reuse have been conducted in Denver; indirect potable reuse via surface sources has been studied in San Diego and Tampa; and indirect reuse via injection has been practiced and studied in Orange County, California. Such studies employ state-of-the-art methodologies to measure toxicological effects and determine the identities of inorganic and organic chemical compounds. None of the studies found significant effects from chemical toxicants or infectious disease agents, although methodological limitations and the limited extent of testing prevent us from interpreting these results as showing with complete certainty that there are no health effects associated with human consumption of recharged water from impaired quality sources, especially over the long term. Thus, while studies to date fail to show that professionally managed recharge projects produce extracted water of lower quality from a health perspective than water from historically acceptable sources, the methodologies used in such assessments have limitations and, accordingly, uncertainties remain. Also, although various treatment technologies are available, there is always some inherent uncertainty and risk of failure because of the possibility of human error or equipment malfunction.

The information available from on-site and laboratory studies do not indicate that the health risks from recovered water are greater than those from existing water supplies or that the concentrations of chemicals or microorganisms are likely to be higher than those established in drinking water standards set by the Environmental Protection Agency (EPA). There are uncertainties, however, such as limited chemical and toxicological characterizations of source waters, the absence of water quality standards for some of the chemicals, and the uncertain environmental fates of chemicals and microorganisms in the recharge system. Furthermore, it should be remembered that the EPA standards are based on water sampled from high quality sources. Such standards are often years behind current knowledge and current knowledge at any time is limited. (For example, health effects are determined for each organic compound separately, not for the inevitable mixtures of organics.) Accordingly, monitoring of potentially toxic constituents and pathogenic microorganisms should be required in using water extracted from recharge systems.

ECONOMIC, LEGAL, AND INSTITUTIONAL CONSIDERATIONS

The future of artificial recharge using waters of impaired quality will be crucially affected by the economic, legal, and institutional setting. Indeed, the institutional barriers may prove to be more problematic than the remaining technical constraints.

From an economic perspective, recharge with waters of impaired quality may be a more attractive option in the future both because of the increasing scarcity of new surface water sources and because of increasingly stringent wastewater discharge regulations. Although the economic feasibility of recharge varies from site to site, in general, recharge will be economically attractive whenever it is the least cost source of supplemental water. The cost of treatment over and above what is required to meet wastewater discharge standards will be particularly important. Cost is, of course, sensitive to the distance the recovered waters must be transported for spreading or injection and to the recharge techniques used. Economic feasibility also will depend on the benefits ultimately provided by the recovered water.

From both an economic and a legal perspective, the need to define rights to both source waters and recovered waters is paramount. Failure to define water rights clearly makes recharge with waters of impaired quality a far less attractive option than otherwise might be the case. From a strictly legal standpoint, the central question is how to formulate policy to protect public health and the environment, while not imposing inappropriate or unnecessarily burdensome controls on this potentially important form of water development. State laws governing recharge vary greatly. California is developing a comprehensive set of laws and regulations related to recharge, but most states have not addressed this regulatory problem adequately. Although there are federal laws that govern certain aspects of the recharge process, the federal government has not exercised strong leadership in developing appropriate institutions to govern wastewater discharge and reuse.

Another potential constraint is public perception. Indeed, the importance of the attitudes of the public cannot be discounted because the public is ultimately the recipient of the recovered water and it ultimately, albeit often indirectly, bears the burden of the costs of such operations. People's attitudes about the reuse of water in general depend on the source and the intended purpose of the reuse; nonpotable reuse is relatively well accepted, but potable reuse and other high-contact uses are not favored because of perceived health risks and water quality problems. Where artificial recharge is planned, especially where impaired quality sources are used, early public involvement can help develop the public's understanding of the issues.

CONCLUSIONS

As demand for water increases, water managers and planners will need to look widely for ways to improve water management and augment water supplies. The Committee on Ground Water Recharge concludes that artificial recharge can be one option in an integrated strategy to optimize total water resource management, and it believes that with pretreatment, soil-aquifer treatment, and posttreatment as appropriate for the source and site, impaired-quality water can be used as a source for artificial recharge of ground water aquifers.

Artificial recharge using source waters of impaired quality is a sound option where recharge is intended to control saltwater intrusion, reduce land subsidence, maintain stream baseflows, or similar in-ground functions. It is particularly well-suited for nonpotable purposes, such as landscape irrigation, because health risks are minimal and public acceptance is high. Where the recharged water is to be used for potable purposes, the health risks and uncertainties are greater. In the past, the development of potable supplies has been guided by the principle that water supply should be taken from the most desirable source feasible, and the rationale for this dictate remains valid. Thus, although indirect potable reuse occurs throughout the nation and world wherever treated wastewater is discharged into a water course or underground and withdrawn downstream or downgradient for potable purposes, such sources are in general less desirable than using a higher quality source for potable purposes. However, when higher-quality, economically feasible sources are unavailable or insufficient, artificially recharged ground water may be an alternative for potable use.

The following recommendations emerged from the committee's deliberations; an expanded discussion of these points appears in chapter 7.

Artificial Recharge: A Viable Option

Artificial recharge of ground water using source waters of impaired quality can be a viable way to augment regional water supplies—primarily for nonpotable uses but sometimes for potable uses under appropriate conditions—and at the same time provide an avenue for wastewater management.

- Once recharge has been deemed feasible as part of an integrated approach to regional water supply planning, the method of recharge chosen should be based on hydrogeologic conditions and the specific benefits sought from the recharge. In general, surface spreading offers the greatest engineering and operational advantages. Surface methods can accommodate waters of poorer quality and are simpler to design and operate than recharge wells, although certain conditions may require use of wells. Because surface spreading requires large amounts of land with permeable soil, it may not be feasible in densely populated areas or elsewhere where suitable land is expensive or unavailable. Injection wells require high quality source water to avoid clogging problems and also because aquifers alone do not provide the same degree of treatment as soil aquifer systems. Although there are indications of some water quality improvements within aquifers, considerable pretreatment is necessary if the source water to be used in wells is of impaired quality.

- Artificial recharge using water of impaired quality offers particularly significant potential for nonpotable uses. Nonpotable reuse can help reduce demand on limited fresh water sources at minimal health risk; it is widely practiced and achieves good public acceptance. Potable reuse is equally possible to engineer, but the health risks may be greater and public acceptance is less certain. In either approach, but especially where potable reuse is considered, careful preproject study and planning is required.

Potential Impaired Quality Sources

Three main types of impaired quality waters are potentially available for ground water recharge—treated municipal wastewater, stormwater runoff, and irrigation return flow. Of these, treated municipal wastewater is usually the most consistent in terms of quality and availability. Stormwater runoff from residential areas generally is of acceptable quality for most recharge operations, but at some times and places it may be heavily contaminated, and its availability is variable and unpredictable. Irrigation return flow exhibits wide variations in quality and is sometimes seriously contaminated, and thus usually is not a desirable source of water for recharge.

- Based on current information, treated municipal wastewater intended as a source for artificial recharge should receive at least secondary treatment. Municipal wastewater that has received only primary treatment may be adequate for the recharge of nonpotable ground water in certain areas, but use of primary effluent should not be considered without implementation of a site-specific demonstration study.

- Certain impaired quality waters, such as irrigation return flow, stormwater runoff from industrial areas, and industrial wastewater, generally should not be regarded as suitable sources for artificial recharge. Exceptions might be identified, but only after careful characterization of source water quality on a case-by-case basis. Other types of stormwater runoff to avoid include most dry weather storm drainage flow, salt-laden snowmelt flow, and flow originating from certain commercial facilities, such as vehicle service areas. Construction site runoff also should be avoided to prevent clogging of recharge facilities with eroded soil and other debris.

Human Health Concerns

The principal concern with regard to artificial recharge using waters of impaired quality for potable purposes is the protection of human health. Several major studies employing state-of-the-art methods for organic analysis and toxicological testing show that well managed recharge projects produce recovered water of essentially the same quality from a health perspective as water from other acceptable sources. However, there are uncertainties in identifying potentially toxic constituents and pathogenic agents in the methodologies used in these studies, and thus potable reuse should be considered only when better quality sources are unavailable.

- Disinfection of treated municipal wastewater prior to recharge should be managed so as to minimize the formation of disinfection by-products. Alternatives to chlorination include disinfection with ultraviolet radiation and the use of other chemical disinfectants. However, additional research should be undertaken on pathogen removal, formation of disinfection by-products, and removal of disinfection by-products before alternative disinfectants can be classified as conclusively superior to chlorine.
- Recovered water must be monitored carefully to provide assurance that pathogenic microorganisms and toxic chemicals do not occur at concentrations that might exceed drinking water standards or other water quality parameters established specifically for reclaimed water which consider the nature of the source water. The outcomes of existing studies of potable use of recovered ground water recharged with treated municipal wastewater suggest that additional epidemiological, in-vivo, or short-term toxicological studies would be of marginal value. As long as the recovered water meets drinking water standards and other water quality limits specified for the site, and there is no evidence from

monitoring of constituents that pose undesirable health risks, additional toxicological testing is unnecessary. If the quality of the extracted water is uncertain for any reason, it should not be considered for potable reuse.

• There are significant uncertainties associated with the transport and fate of viruses in recharged aquifers. These uncertainties make it difficult to determine the levels of risk of any infectious agents still contained in the disinfected wastewater. Thus, additional research should be undertaken on the transport and fate of viruses in recharged aquifers to allow improved assessments of the possible health risks and needs for post-extraction disinfection associated with such systems.

• Artificial recharge of ground water with waters of impaired quality should be used to augment water supplies for potable uses only when better-quality sources are not available, subject to thorough consideration of health effects and depending on economic and practical considerations.

System Management and Monitoring

Protecting public health and the sustainability of soil-aquifer systems will require careful planning, operation, and management of recharge systems. Under appropriate conditions, the soil-aquifer system has the capacity to remove certain chemicals and pathogens and can therefore be an effective component in ground water recharge and water reuse systems. However, the processes through which removal occurs are not completely efficient in natural settings, and not all constituents are retained or degraded to the same extent. In addition, strategies that may enhance the removal of one chemical or pathogen can decrease the efficiency of removal of another.

• Assessments of the feasibility of any recharge technology should include analyses of the possible impacts of the use of the system on the environment.

• Monitoring of recharge water should be undertaken as it moves toward points of recovery. This is critical to help ensure that water quality is maintained, to provide early warning of unexpected problems, and to help maintain the long-term viability of the treatment system.

Economic Considerations

Artificial recharge opportunities need to be evaluated within the overall context of available water supplies, existing and projected water demands, and related costs and benefits to ensure that the opportunity is economically justified.

• The price of recovered water should reflect the true cost of making the water available to ensure that the water is used efficiently. The cost of recharge operations should not be subsidized to make this water source more attractive than it would otherwise be.

Legal and Institutional Considerations

The development of institutional arrangements governing artificial recharge is critical in determining the extent to which water supplies will ultimately be available from recharge with waters of impaired quality. Institutions need to be capable of formulating policies to protect public health and safety and environmental amenities while not imposing inappropriate or inefficient controls on this potentially important form of water resource management. Federal leadership will be needed if the full promise of artificial ground water recharge is to be realized.

- As a first step in developing institutional arrangements that will foster artificial recharge as a means of augmenting water supplies, states should move to clarify the legal rights to source waters and recovered waters for artificial recharge operations.
- In addition to ensuring the protection of public health related to the consumption of recovered water, when developing regulatory policies states should make explicit provision for the evaluation of project sustainability and environmental impacts of artificial recharge projects.
- Regulatory processes should ensure that environmental impacts and other third party effects are adequately accounted for in the design and operation of artificial recharge projects.
- The federal government should assume leadership in supporting the development of artificial recharge with treated municipal wastewater and other suitable impaired-quality water sources by providing technical assistance to the states and by developing model statutes and guidelines.

REFERENCES

Bouwer, H. 1978. Ground Water Hydrology. New York: McGraw Hill. 480 pp.
Bull, R. J., C. Gerba, and R. R. Trussell. 1990. Evaluation of the health risks associated with disinfection. Critical Reviews in Environmental Control 20:77-113.

1

An Introduction to Artificial Recharge

One result of the growing competition for water is increased attention to the use of artificial recharge to augment ground water supplies. Stated simply, artificial recharge is a process by which excess surface water is directed into the ground—either by spreading on the surface, by using recharge wells, or by altering natural conditions to increase infiltration—to replenish an aquifer. Artificial recharge (sometimes called planned recharge) is a way to store water underground in times of water surplus to meet demand in times of shortage. Water recovered from recharge projects can be allocated to nonpotable uses such as landscape irrigation or, less commonly, to potable use. Artificial recharge can also be used to control seawater intrusion in coastal aquifers, control land subsidence caused by declining ground water levels, maintain base flow in some streams, and raise water levels to reduce the cost of ground water pumping.

It is useful to think of the entire artificial recharge operation as a water source undergoing a series of treatment steps during which its composition changes. The constituents of potential concern depend not only on the character of the source water, but also on its treatment prior to recharge (pretreatment), changes that occur as it moves through the soil and aquifer (soil-aquifer processes), and treatment after withdrawal for use (posttreatment).

This report discusses three types of source waters having very different characteristics—treated municipal wastewater, stormwater runoff, and irrigation return flow—that have been proposed for use in artificial recharge. Normally, each of these source waters needs to be subjected to some kind of pretreatment before being introduced into the soil or aquifer. The exact pretreatment operations required depend on the type of source water, the nature of the recharge process, and the intended use of the recovered water. A fundamental assumption of this report is that wastewater used to recharge the ground water must receive a sufficiently high degree of treatment prior to recharge so as to minimize the extent of any degradation of native ground water quality, as well as to minimize the need for and extent of additional treatment at the point of extraction.

After pretreatment, the water is ready for recharge, either through surface spreading and infiltration through the unsaturated zone or by direct injection into ground water. Recharge by infiltration takes advantage of the natural treatment processes, such as biodegradation of organic chemicals, that occur as water moves through soil. The quality of the water prior to recharge is of interest in assessing the possible risks associated with human exposures to chemical toxicants and pathogenic microorganisms that might be present in the source water. Although one can reasonably expect that such constituents will often be reduced during filtration through the soil, as well as subsequently in the aquifer, a conservative approach to risk assessment would assume that toxicants and microorganisms are not completely removed and some are affected only minimally prior to subsequent extraction and use. Thus when recharge water is withdrawn later for another purpose, it may require some degree of posttreatment, depending on its intended use.

Taking a systems perspective that encompasses all steps from pretreatment, through recharge, through transformation and transport, to extraction, this report assesses the issues and uncertainties associated with the artificial recharge of ground water using source waters of impaired quality. In particular, the report focuses on the methodologies and nature of the recharge systems and the subsequent impacts on the native ground water quality, especially as those impacts may affect public health following use of the recovered water. Economic, institutional, and regulatory questions are examined as well. First, this chapter presents a primer on artificial recharge of ground water to give the reader an introduction to the philosophy and techniques of the field.

A PRIMER ON ARTIFICIAL RECHARGE

Water continually evaporates from the oceans and other open water bodies, moves across the land as water vapor in clouds, falls back on the land as rain and snow, and then returns to the oceans through rivers and underground pathways to start the cycle—the hydrologic cycle—again. Part of the water that falls on the land evaporates from the soil or is transpired from plants back into the atmosphere. Another part flows overland to stream channels, lakes, or the sea. The remainder seeps downward through the soil under the influence of gravity to enter the ground water system. Once in the ground water system, the water moves slowly in response to ground water slopes or hydraulic gradients until it reenters the surface part of the cycle.

The term ground water applies to water of higher than atmospheric pressure contained below the land surface in saturated fractures, cracks, cavities, and pore spaces in geologic formations. It is distinct from water in the unsaturated zone, which can be at or below atmospheric pressure and is contained in films and pores in the partially air-filled soil region between the ground water zone and the soil surface. This upper region containing soil, water, and air is called the unsaturated, or vadose, zone. The term recharge is used for water entering the ground water system and the term discharge applies to water leaving it. Geologic units permeable enough to yield appreciable amounts of ground water to wells are termed aquifers.

Ground Water Flow

The upper surface or boundary of the zone of complete saturation in the ground water system is called the water table. In general, the water table stands higher under hills than under valleys, and these differences in water table height (also referred to as head differences) provide the hydraulic gradients that cause ground water to flow from recharge areas to areas of discharge. Ground water in the shallow geologic units containing the water table is called unconfined because the water is more or less in direct contact with the atmosphere.

Recharge to a water table aquifer occurs wherever rainfall or surface water infiltrates downward through the soil to the water table. Generally, the recharge area of an aquifer is the entire land surface overlying the aquifer, although certain portions, such as those lying under lakes or streams, may supply much of the recharge volume. Ground water in deeper geologic units separated from the water table beds by confining layers is said to be under confined, or artesian, conditions. The height of the water level in a well open to a confined aquifer marks the position of the potentiometric surface of that aquifer. Where the height of the potentiometric surface is higher than that of the land surface, wells open to the confined aquifer will flow freely (Figure 1.1).

Recharge to a confined aquifer can occur if the pressure of its water is less than that in the overlying or underlying aquifers that adjoin it. Discharge of ground water takes place through springs, streams, wetlands, lakes, tidal waters, and pumped wells. Water tables and potentiometric surfaces fluctuate seasonally, generally by several feet, in response to natural variations in rates of recharge and discharge.

Ground water is constantly in motion, following hydraulic gradients from points of high head to points of low head in an aquifer system. Flow of ground water is always laminar, except near large springs or pumped wells, where it may be turbulent. The velocity of ground water flow in aquifers generally ranges from a few inches to a few feet per day and is determined by the porosity, permeability, and hydraulic gradient.

Ground water naturally contains concentrations of various mineral substances that have been dissolved from the local soil and geologic formations. In

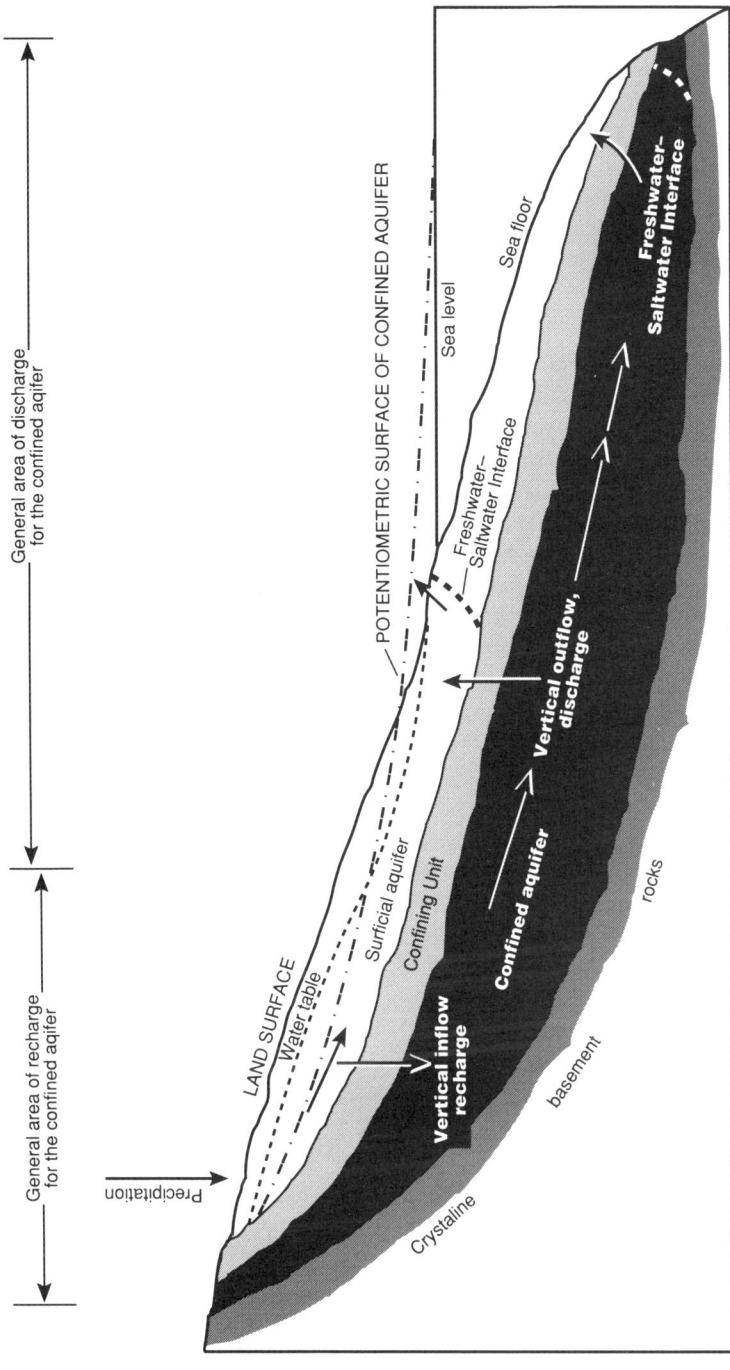

FIGURE 1.1 Idealized hydrogeologic cross section showing features of the ground water flow system. Source: Modified from Giese, G. L., J. L. Eimers, and R. W. Coble. 1991. Simulation of ground-water flow in the coastal plain aquifer system of North Carolina: U.S. Geological Survey Open-File Report 90-372, Figure 4, p. 11.

general, the longer the water remains in the earth and the higher its temperature, the higher the concentrations of dissolved substances. Thus, deep ground waters generally tend to contain more dissolved salts than shallow ground waters. Disposal of man-made wastes adds additional substances to the ground water, sometimes degrading the quality of the water so that it no longer is potable. In coastal areas, the inland extent of salty ground water is controlled by heads in the fresh ground water system. Lowering of those heads by pumping induces saltwater intrusion.

Artificial Recharge

Artificial recharge is the process of spreading or impounding water on the land to increase the infiltration through the soil and percolation to the aquifer or of injecting water by wells directly into the aquifer. Surface infiltration systems can be used to recharge unconfined aquifers only. Confined aquifers can be recharged with wells that penetrate the aquifer. Well recharge is also used for unconfined aquifers if suitable land for infiltration systems is not available.

Artificial recharge can be done using any surplus surface water. When low-quality water is used for recharge, the underground formations can act as natural filters to remove many physical, biological, and chemical pollutants from the water as it moves through. Often, the quality improvement of the water is actually the main objective of recharge, and the system is operated specifically using the soil and the aquifer to provide additional treatment to the source water. Systems used in this way are called soil-aquifer treatment (SAT), or geopurification, systems.

The water extracted from SAT systems often can be used without further treatment to support recreation, landscape irrigation, and other nonpotable purposes. Potable use may require more treatment. Because aquifers usually are much coarser than vadose zones, the quality improvement of the water is much less in the aquifer than in the vadose zone. Thus, recharge using wells in confined aquifers cannot be expected to produce major improvements in the quality of the water. If low-quality water is to be used for well injection, it must be treated to meet the desired reuse qualities before injection. In addition, adequate treatment of the water before recharge is necessary to reduce clogging of the recharge wells. An overview of sources of water, treatment options, recharge systems, recovery techniques, and uses of the water after recovery is given in Table 1.1.

Surface Infiltration Systems

Surface infiltration systems designed to provide artificial recharge of ground water require permeable soils (sandy loams, sands, gravels) that have relatively high infiltration rates and that can transmit the applied water without completely

TABLE 1.1 Overview of Impaired Quality Water Sources for Artificial Recharge, Treatment Options, Recharge Systems, and Uses of the Water After Recovery

Possible Sources of Impaired Quality Water
- Treated municipal wastewater
- Stormwater runoff
- Irrigation return flow

Treatment Options Before Recharge
- None
- Primary (sedimentation)
- Filtration
- Secondary
- Tertiary (secondary and filtration)
- Disinfection
- Advanced (removal of metals, nitrogen, phosphorus, total organic carbon, and total dissolved solids)

Recharge Systems[a]
- Surface infiltration (provides maximum SAT)
- Dry wells (provides some SAT)
- Injection recharge (provides minimal SAT)

Approach to Recovery from Aquifer
- Systematic (100% recovery)
- Random (dilution with native ground water)
- Pumped wells
- Gravity drains
- Natural drainage into surface water

Treatment Options After Recovery
- None
- Disinfection
- Membrane filtration (reverse osmosis)
- Activated carbon filtration

Use of Recovered Water
- Nonpotable
    unrestricted irrigation (vegetables, playgrounds)
    unrestricted recreation (streams, lakes)
    industrial (cooling, processing, construction)
    fire fighting
    toilet flushing
    environmental (in-stream benefits, wildlife refuges, wetlands)
- Potable (after dilution and/or posttreatment)

[a]SAT = soil-aquifer treatment.

FIGURE 1.2 T-levees for spreading water in the Santa Ana River, California. Except for some storm runoff events, the water is almost all treated municipal wastewater from inland cities. Credit: H. Bouwer, U.S. Water Conservation Laboratory, Phoenix, Arizona.

saturating the zone above the ground water. Conventional infiltration systems can be grouped into in-channel and off-channel systems. In-channel systems are weirs, dams, or T- or L-shaped levees that spread the water over a streambed or floodplain (Figure 1.2). Dams must be built with adequate spillways or washout sections to handle spring runoff or other periodic large flows. Inflatable rubber dams that are deflated to pass large flows also can be used. The smaller weirs and levees are often considered expendable and are easily reconstructed after damage by high flows. Off-channel systems may consist of old gravel pits or of specially built basins (Figure 1.3). In-channel and off-channel infiltration systems are common in California, where there are a large number of successful recharge projects.

The range of infiltration rates (i.e., the rate water drains into the ground when a basin is flooded) for in- and off-channel systems is about 0.3 to 3 meters

AN INTRODUCTION TO ARTIFICIAL RECHARGE

FIGURE 1.3 Infiltration basin near Palm Springs, California, using Colorado River water. The windmills are for power generation. Credit: H. Bouwer, U.S. Water Conservation Laboratory, Phoenix, Arizona.

(m) (1 to 10 ft) per day, including any effects caused by clogging (Bouwer and Rice, 1984). Systems with year-round recharge and periodic drying and cleaning of the bottom typically have hydraulic loading rates of 30 to 300 m/year (98 to 980 ft/year). Evaporation rates from water surfaces and wet soils range from 0.3 m/year (1 ft/year) or less in cool, humid climates to 2.5 m/year (8.2 ft/year) or more in warm, dry climates. Thus, evaporation losses are much less than the amounts that infiltrate into the ground. If the basin bottoms are not covered by sediment or other clogging material and ground water levels are sufficiently low to not affect infiltration, infiltration rates are about the same as the vertical hydraulic conductivity of the soil, which may be about 0.3 m/day (1 ft/day) for sandy loams, 1 m/day (3.3 ft/day) for loamy sands, 5 m/day (16 ft/day) for fine sands, 10 m/day (33 ft/day) for coarser sands, and 20 to 50 m/day (66 to 160 ft/day) for fine or clean gravel. Sand and gravel mixtures have lower hydraulic conductivities than the sand alone (Bouwer and Rice, 1984). To achieve optimal infiltration rates, a number of features need to be considered in the design process, including clogging, water depth, and ground water level.

Clogging

A major operational feature of infiltration systems for artificial recharge of ground water is soil clogging caused by to accumulation of suspended solids on the bottom and banks of the infiltration facility as they settle or are strained out on the soil surface. The suspended solids can be inorganic (e.g., clays, silts, fine sands) or organic (e.g., algae, bacterial flocks, sludge particles). Also, biofilms can grow on the bottom. Some mobile bacteria actually may produce mats of polymer strands, which can then strain out fine suspended particles. Thus, clogging layers may consist of a mixture of organic and inorganic products. Their thickness may range from 1 millimeter (mm) (0.039 inch) or less to 0.3 m (1 ft) or more. They have a low permeability and, hence, they reduce infiltration rates (Bouwer, 1982).

As the clogging layer forms and its hydraulic resistance increases, the clogging layer becomes the controlling factor of the infiltration process, and infiltration rates decrease. When infiltration rates become unacceptably low, the infiltration system must be dried to restore infiltration rates. If the clogging layer is primarily organic, as, for example with treated municipal wastewater, drying alone may be sufficient to restore infiltration rates. The clogging layer then partly decomposes, cracks, and curls up to form flakes on the bottom. When the basins are flooded again, essentially normal infiltration rates are obtained until the clogging process repeats itself. The problem, then, is to find the optimal combination of flooding and drying periods that yields maximum long-term accumulated infiltration rates.

Because of the many variables involved, such optimal combinations are best found by site-specific experimentation. For treated municipal wastewater with a low suspended solids content (for example, less than 10 milligrams per liter (mg/l)), such combinations typically range from 2 days of flooding and 5 days of drying to 2 weeks of flooding and 2 weeks of drying. Hydraulic loading rates (i.e., average infiltration over time, including wet, dry, and cleaning cycles) then are typically in the range of 30 to 200 m/year (98 to 660 ft/year), depending on effluent quality, soil, and climate (Bouwer, 1982). Flooding and drying schedules may also be controlled by environmental factors such as breeding of insects, protection of wildlife, formation of floating algae, odors, and recreational uses of the infiltration system.

As clogging material continues to accumulate on the bottom and banks of the infiltration facilities, it eventually reduces infiltration rates so much that it should be removed. For treated municipal wastewater this process may have to be repeated every 1 or 2 years if the municipal wastewater has had adequate pretreatment and clarification (suspended solids contents less than 10 mg/l). On the other hand, if wastewater with a very high suspended solids content is used the clogging layer may have to be removed at the end of every drying period.

Clogging layers promote unsaturated flow in the vadose zone, and they are active biofilters that can remove fine suspended solids, microorganisms, organic carbon, nitrate, and metals from the water as it moves through them. This property can be of great importance in wastewater lagoons or constructed wetlands where underlying ground water needs to be protected against pollution. Clogging layers may also be desirable for infiltration systems in very coarse sands to reduce infiltration rates and enhance SAT benefits if water of low quality is used.

Water Depth

The water depth in infiltration basins is selected carefully. While high hydraulic heads produced by deep water result in high infiltration rates, they also tend to compress clogging layers. Thus, contrary to intuitive expectations, deep basins can produce lower infiltration rates than shallow basins (Bouwer and Rice, 1989). Also, the rate of turnover of the water in deep basins may be less than in shallow basins, allowing suspended algae (for example, *Carteria klebsii*) to grow in longer exposure to sunlight. Algae causes additional clogging of the soil as the biomass is strained out by infiltration. Another undesirable effect is that calcium carbonate may precipitate because of increases in the pH of the water as photosynthesizing algae take up dissolved carbon from the water, further aggravating the clogging.

In addition to yielding higher infiltration rates, shallow basins (water depths about 20 centimeters (cm) (8 inches) or less) have the advantage that drying can start quickly (in less than 24 hours, for example) after the inflow of water into the basin has been stopped. The water then disappears quickly by infiltration, or it can be drained into a lower basin. With deep basins (for example, 10 m (33 ft) or more, as in old gravel pits), it can take a very long time for all the water to infiltrate into the soil. Water may actually have to be pumped out of the basin to initiate drying. On the other hand, shallow basins may have more weed growth, but this can be controlled.

Ground Water Level

Another design criterion is that the ground water table must be deep enough below the infiltration system that it does not interfere with the infiltration process. This requirement applies to the mounding of the permanent water table caused by recharging, as well as to perched ground water mounds that may form over restricting layers in the vadose zone. Where infiltration rates are controlled by the clogging layer (which is the rule rather than the exception for basins and ponds), the water table must be at least 0.5 m (1.6 ft) below the bottom of the basin. This distance usually is adequate to keep the top of the capillary fringe below the basin bottom, so that infiltration rates are not restricted by underlying ground water.

Where there is no clogging layer, there is more hydraulic continuity between the water in the infiltration system and the ground water. In that case, the vertical distance between the water surface and the ground water table (at some distance from the ponds where most of the mound has dissipated) should be at least twice the width of the infiltration system (Bouwer, 1990). Thus, where ground water levels are high, maximum infiltration rates can be obtained only with long, narrow streams or basins spaced a suitable distance apart.

Where waters of impaired quality are used for recharge by surface infiltration systems, it may be desirable to keep ground water levels sufficiently low to create an adequate unsaturated zone below basin bottoms for aerobic processes and virus removal. Proposed California regulations, for example, require a minimum depth to ground water of 3 m (9.8 ft) below the basins (Hultquist et al., 1991). Other infiltration systems, however, such as those in the dunes of The Netherlands for pretreatment of Rhine water for potable use, operate essentially in the ground water zone with no unsaturated conditions. Also, where wells are drilled close to streams or lakes to "pull" surface water through the aquifer for treatment prior to drinking water treatment (bank filtration systems), the processes also take place completely below the ground water table. Thus, there is no standard for minimum depth to ground water below infiltration basins for adequate quality improvement of waters of impaired quality.

Water Quality

Impaired quality water sources can vary in quality. For relatively unpolluted water, the most important quality parameters applicable when considering ground water recharge are suspended solids (SS) content, total dissolved solids

(TDS) content, and the concentrations of major cations such as calcium, magnesium, and sodium. As noted earlier, suspended solids cause clogging of bottoms and banks. Where the SS content is too high, presedimentation basins with possible use of coagulants may be required. On-site experimentation will determine the best combination of pretreatment and cleaning schedules for maximum hydraulic capacity and economy of operation.

High concentrations of sodium ions in soil water, particularly in water of overall low salt concentration, can break apart aggregated clay and liberate individual colloids that subsequently plug soil pores and reduce hydraulic conductivity and infiltration rate. TDS levels and concentrations of calcium, magnesium, and sodium (as reflected by the sodium adsorption ratio (SAR)) are used to determine whether a water will disperse or flocculate clay (Bouwer, 1978). Because of effects on status of clay in the clogging layer, for example, recharge basins in California typically produce higher infiltration rates for Colorado River water (high TDS, low SAR) than for Sacramento River water from the California aqueduct (lower TDS, higher SAR). Thus, TDS, calcium, and magnesium contents should be high enough and sodium content low enough in the recharge water to keep clay in the clogging layer and below in a flocculated, more permeable state. Aquifers sometimes contain clay lenses. If the recharge water has a low TDS and/or a high sodium concentration, this clay could then become dispersed and move with the ground water through the coarse layers of the aquifer, causing the water pumped from wells in the aquifer to be "muddy."

Water quality issues ultimately affect the intended use of the recharge water and whether it is suitable for potable or nonpotable purposes. When potable uses are planned, the parameters of concern include microorganisms, trace inorganic chemicals, and anthropogenic organic chemicals including disinfection by-products. These constituents and issues related to them are examined at length in Chapter 4 in this report.

Soil-Aquifer Treatment

Where treated municipal wastewater or stormwater runoff are used in surface infiltration systems, the vadose zone and in some cases the aquifer act as natural, slow filters that typically reduce the concentration of various pollutants due to physical, chemical, and microbiological processes. Suspended solids are filtered out; biodegradable organic compounds are decomposed; microorganisms are adsorbed, strained out, or die because of competition with other soil microorganisms; nitrogen concentrations are reduced by denitrification; synthetic organic compounds are adsorbed and/or biodegraded; and phosphorous, fluoride, and heavy metals are adsorbed, precipitated, or otherwise immobilized. Thus, soil-aquifer treatment can be an important step in the treatment train for reuse of wastewater.

Most soil-aquifer treatment processes take place in the upper part of the vadose zone where soils generally are finer and have a greater organic matter content than in the aquifer, the flow is unsaturated, and oxygen levels vary from aerobic to anaerobic. To protect high-quality native ground water and nearby drinking water wells against encroachment by sewage-derived water or recharge water of other impaired quality, the systems normally are designed as recharge-recovery systems, where all the recharge water is taken out of the aquifer again with strategically located wells, drains, or other interceptors (Figure 1.4).

If the recovered water is 100 percent treated municipal wastewater, as is possible in systems B and C in Figure 1.4 (assuming for system B that there are also infiltration basins on the other side of the drain), the recovered water can be treated further (posttreatment) to meet the quality requirements for the intended use. This approach permits selection of the most economical treatment train of pretreatment, SAT, and post-treatment to achieve the desired quality of the final product water.

For the United States, the typical treatment train for municipal wastewater might include primary and secondary treatment followed by disinfection, SAT, and no treatment of the withdrawn water if it is to be used for nonpotable purposes. If viruses and other pathogens are found in the water after SAT, it can be disinfected further if it is to be used for unrestricted irrigation (e.g., of crops consumed raw or brought raw into the kitchen, parks, playgrounds, golf courses, private yards) or unrestricted recreation (such as lakes for swimming). Primary treatment alone may be sufficient as pretreatment for municipal wastewater. The higher total organic carbon (TOC) content of primary effluent may enhance the quality improvement gained via SAT because it increases denitrification and biodegradation of TOC. The latter is due to the co-metabolism and secondary utilization brought on by the greater availability of organic carbon (McCarty et al., 1984). The level of pretreatment needed is highly site- and use-specific and can be selected only after careful study.

Where the water after SAT is to be used for drinking, posttreatment may be necessary to remove residual TOC and possibly pathogens that have survived SAT. The treatment could include activated carbon filtration, reverse osmosis or other membrane filtration, and disinfection. Also, more pretreatment may be done (e.g., nitrogen removal, filtration, and disinfection) where required by local regulations, where soils and aquifers are too coarse to provide adequate treatment, or where the water after SAT is pumped from a random or nonsystematic layout of wells, which makes treatment after SAT unfeasible.

Sometimes, dilution with native ground water is relied on to allow potable use of the recovered water without further treatment. Proposed California regulations, for example, require that well water from SAT systems using treated municipal wastewater consist of not more than 20 or 50 percent sewage (depending on the level of pretreatment and site conditions) (Hultquist et al., 1991).

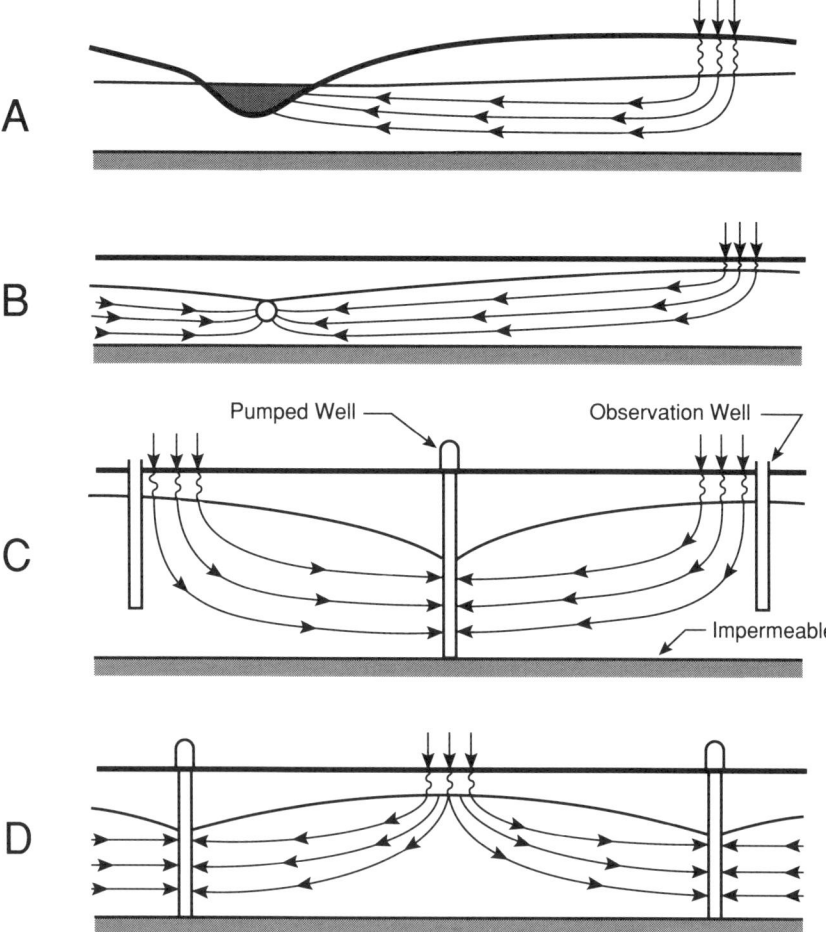

FIGURE 1.4 Schematic showing lines of flow in recharge - recovery SAT systems with (A) natural drainage of renovated water into stream, lake, or low area, (B) collection of renovated water by subsurface drain, (C) infiltration areas in two parallel rows and line of wells midway between, and (D) infiltration areas in center surrounded by a circle of wells. Source: H. Bouwer.

System Design and Pilot Projects

Infiltration systems for artificial recharge of ground water or SAT systems for treatment and storage of waters of impaired quality must be tailored to local hydrogeology, quality of input water, and climate. In general, basin water depths should be less than 30 cm (1 ft), and the basins should be hydraulically independent so that each can be flooded, dried, and cleaned according to its best schedule. Inlet structures must not cause soil erosion that could clog basin bottoms. Drying periods should be started before infiltration rates have reached low values so that drying can be achieved by natural infiltration, and pumping or draining the basins is not necessary. There should also be a sufficiently large number of basins to permit flexible operation (variable flooding, drying, and cleaning cycles), with some basins in reserve to handle maximum water flows or flows during periods of low infiltration rates. Rates can be low, for example, in the winter when the water is cold, drying is slow, and infiltration recovery is incomplete, or in the summer when algae and bottom biofilms grow faster.

Where there is no local experience with artificial recharge, adequate site investigations and local experimentation with a pilot or test project are necessary, especially if the source water is treated municipal wastewater or other low-quality water. The results from such pilot projects are then used to develop design and management criteria for optimal performance of the full-scale system. Even when the full-scale system is constructed and in operation, fine-tuning may be necessary to improve its performance. Finally, infiltration systems must be closely managed to monitor their performance and to allow for quick action when something goes wrong.

When recharge systems using treated municipal wastewater are underdesigned or not properly managed (e.g., not dried and cleaned frequently enough), their infiltration rates go down and operators eventually have to fill all the basins with no time for drying. This situation causes further declines in infiltration rates while the wastewater keeps coming in and water depths in the basins increase. This compresses the clogging layer and causes more growth of algae in the basins, and infiltration rates are further reduced until eventually the whole system is ineffective. To prevent this reduction in efficiency, there must always be sufficient basin area to maintain shallow water depths and to allow regular drying and cleaning for maintaining infiltration rates.

Well Injection Systems

Ground water recharge with surface infiltration systems is not feasible where permeable surface soils are not available, land is too costly, vadose zones have restricting layers or undesirable natural or synthetic chemicals that can leach out, or aquifers have poor-quality water at the top or are confined. For those conditions, ground water recharge with recharge wells is an option. These wells are similar to regular pumping wells. For unconsolidated aquifers (sand, gravel), they consist of a casing, screen, gravel pack, grouting, and a pipe to apply water to the well for infiltration into the aquifer (Figure 1.5). For consolidated aquifers (sandstone, fractured rock, limestone with secondary porosity) the portion of the well in the rock is completed as an open borehole without screen or envelope.

AN INTRODUCTION TO ARTIFICIAL RECHARGE

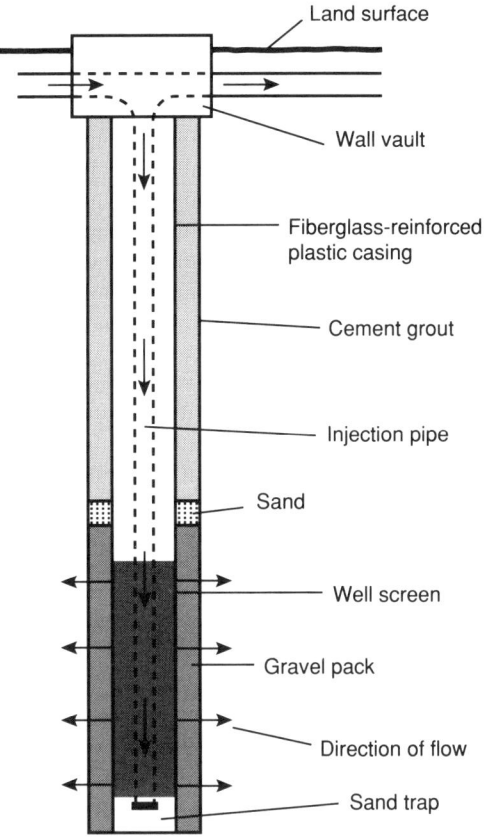

FIGURE 1.5 Schematic of injection well. Source: Modified from Schneider, B. J., H.F.H. Ku, and E. T. Oaksford. 1987. Hydrologic effects of artificial recharge at East Meadow, Long Island, New York. U.S. Geological Survey Water-Resources Investigations Report 85-4323, Figure 8, p. 19.

Some recharge wells have several injection pipes to recharge several confined aquifers.

Beyond their increased costs, the major problem with injection wells is clogging of the aquifer around the well, especially at the borehole interface between gravel envelope and aquifer where suspended solids can accumulate and bacterial growth tends to concentrate. Injection wells are much more vulnerable to clogging than surface infiltrationsystems because the infiltration rates into the aquifer around the borehole are much higher than in infiltration basins. In addition, remediation of clogging in wells is much more difficult than in surface infiltration systems. For injection wells, clogging effects can be remediated by periodic pumping of the wells to reverse the flow and dislodge clogging materials. When recharge wells are pumped, the first water coming out typically is brown and odorous, and must be treated as wastewater or recycled through the water treatment plant. Pumping schedules may range from 20 minutes each day to a few times per year, depending on how fast recharge rates decline. If pumping does not restore recharge rates, redevelopment of the well by surging, jetting, or other conventional well development technique is necessary. Clogging effects can also be overcome by increasing the pressure of the water inside the well. However, increasing the injection pressures too much may actually exacerbate clogging and can cause upward flow of water around the casing or grouting of the well and piping.

The best strategy for dealing with clogging of injection wells is to prevent it by proper treatment of the water before injection. This means removal of suspended solids, assimilable organic carbon, nutrients such as nitrogen and phosphorus, and microorganisms. Also, chlorine is added to maintain a residual chlorine level in the well to minimize microbiological activity. Clogging parameters such as the membrane filtration index, assimilable organic carbon, and clogging in test columns with much higher velocities than in the actual recharge well system (Peters and Castell-Exner, 1993) are useful for identifying relative clogging potentials of various waters. The parameters cannot be used to predict clogging and declines in recharge rates for planned wells because actual clogging often is erratic, seasonal, and sensitive to small changes in water quality and must be evaluated with test wells.

Other processes that can decrease recharge rates in wells are precipitation of calcium carbonate, iron oxides, and other compounds in the aquifer, dispersion and swelling of clay, and air binding. The last can occur when the recharge water is cooler than the aquifer water. As the temperature of the recharge water increases in the aquifer, dissolved air goes out of solution and forms air pockets in the aquifer. These block the pores and reduce the hydraulic conductivity and, hence, the recharge rates. For this reason, dissolved air concentrations in the recharge water should always be as small as possible, and free-falling water in the recharge well should be avoided to prevent air entrainment.

Where municipal wastewater is used for ground water recharge with injection wells, it must undergo extensive pretreatment, including advanced wastewater treatment (AWT) processes. This is necessary to achieve a quality that is essentially the same as that of the drinking water standards before recharge because aquifers generally are too coarse for significant soil-aquifer treatment. The only treatment processes that can be expected in aquifers are some additional TOC removal, removal of some microorganisms, improvement in taste and odor, and similar "aging" and "polishing" effects. In addition, AWT is necessary to minimize clogging in the well. Where the AWT includes reverse osmosis (RO), the water will have a very low TDS concentration, which makes it "hungry" and therefore corrosive. The interaction between this water and the receiving aquifer must then be well understood to make sure that the recharge water does not mobilize undesirable chemicals from minerals and other solid phases of the aquifer. Thus blending of the water after RO with water of a higher TDS content before recharge may be necessary.

Aquifer Storage and Recovery Wells

A rapidly growing practice in artificial recharge is the use of aquifer storage recovery (ASR) wells, which combine recharge and pumping functions. They are used for recharge when surplus water is available, and are pumped when the water is needed. ASR wells typically are used for seasonal storage of drinking water in areas where water demands are significantly greater in summer than in winter, or vice versa. With these wells drinking water treatment plants can be built to meet average, rather than peak, demands.

Dry Wells

Dry wells are boreholes in the vadose zone, usually about 10 to 50 m (33 to 160 ft) deep and about 1 to 1.5 m (3.3 to 4.9 ft) in diameter. They are widely used for infiltration and disposal of stormwater runoff in areas without storm sewers or combined sewers and, hence, they produce incidental recharge of ground water. There is concern that this causes pollution of ground water, but so far studies have not been able to document this. There is also concern about illegal disposal of waste fluids through the dry wells (by so called "night dumpers"). Dry wells normally are drilled into permeable formations in the vadose zone that can accept the runoff water at sufficient rates. Where ground water is relatively deep, dry wells are much cheaper than injection wells and, hence, it is tempting to use dry wells to recharge the ground water instead of injection wells that must go all the way down to the aquifer. To provide adequate recharge, the dry wells should penetrate permeable formations for a substantial distance.

The main problem with dry wells is clogging of the walls. It is impossible to remediate such clogging by pumping or redeveloping because the dry well is in the vadose zone and ground water cannot flow into it. Thus, clogging must be prevented or minimized. One way to do this is to protect the water in the well against the slaking and sloughing of clay layers that could make the water in the well muddy. Slaking can be avoided by filling the well with sand and placing a perforated pipe in the center to carry the water for recharge.

In addition, the water must be pretreated before recharge to remove all clogging agents, including suspended solids, biodegradable organic carbon, nutrients, and microorganisms, and it must be disinfected to maintain a residual chlorine level. If clogging still occurs (and long-term clogging is always a possibility), it will then mostly be caused by bacterial cells and organic metabolic products such as polymers on the well wall (biofouling). Thus, while such clogging cannot be remediated by pumping, cleaning, or redevelopment, a long drying period perhaps could produce significant biodegradation of the clogging material to restore the dry well for recharge.

The choice between using dry wells in the vadose zone or injection wells in the aquifer is governed by economics. The costs of installing the wells, the pretreatment requirements of the water, the useful lives of the wells, and the costs of well replacement and maintenance and remediation need to be compared. If the vadose zone contains undesirable chemicals that can be leached out, dry wells should not be used.

ENVIRONMENTAL EFFECTS

The environmental effects of ground water recharge vary from site to site, and there can be both beneficial and harmful impacts. In general, however, the types of environmental effects that should be considered when planning recharge facilities range from ecological effects on soil, hydrologic, and aquatic ecosystems, to effects on species dependent on riparian habitats, and to possible effects on people's use of the water resources for recreation. As with any use of water, planners must take care to recognize that the impacts of their actions will affect not only local conditions, but also conditions downstream (third-party effects).

The ecological effects of ground water recharge are, for the most part, relatively straightforward and predictable, at least in a qualitative sense. If water is diverted directly from a stream or other surface water source, the reduction in downstream flow will have the same ecological consequences as a diversion for any other purpose that results in a reduction in streamflow with the same timing and quantity. Ecological effects often are difficult to quantify. In almost all cases, they are site specific and difficult to generalize.

During the actual process of recharge, recharge basins produce some significant changes in the ecosystem, both at the surface of the soil and in the soil profile. The nature of these changes depends on whether the recharge basin is a natural stream bed, a lagoon constructed specifically for the purpose, or a natural off-stream basin. The actual management of the recharge basin will often be established so as to manipulate the soil ecology to optimize the recharge process. Bed permeability is always a consideration, and aeration of the bed by intermittent recharge affects the population dynamics of the soil ecosystem. Degradation of organic matter and denitrification are frequent objectives.

The long-term ecological effect on the recharge basin depends on the adsorption of organics and inorganics in the upper soil horizons and on the treatment of the basin when recharge is ceased. Whether the surface topography of the basin is altered to minimize or to maintain inadvertent recharge also can be important.

Downgradient ecological effects depend on the rate of recharge and the nature of the receiving aquifer. If recharge is into a relatively deep aquifer, and if the water table does not approach within a meter of two of the soil surface within the watershed, few ecological consequences are to be expected. On the other hand, if the water table rises anywhere within the basin to within a meter of the soil surface, one can confidently predict sufficient capillary rise to maintain a higher rate of production of vegetation at the surface than would be present otherwise, especially in arid regions. The soil ecology would be similarly affected. Again, the results will be highly site specific, and whether the consequences are considered desirable or undesirable depends on the nature of the effect and the objectives of the recharge. Normally, losses due to evapotranspiration would be regarded as counterproductive because of the loss of stored water and the increase in ground water salinity, which would subtract from other uses of the water.

Along the banks of many rivers in the western United States, the natural recharge results in substantial growth of phreatophytes, such as mesquite. Herbicides, cutting, and other methods to reduce plant growth have been used in attempts to increase streamflow. Although vegetation removal makes more water available, it usually produces a loss rather than a gain in terms of ground water storage because the water moves downstream quickly with little time for infiltration. The removal of vegetation and subsequent increased flow also facilitates erosion. Because the value of riparian habitat, especially in dry regions, is increasingly recognized, many experts believe that the harm of vegetation removal outweighs any potential benefits.

Beyond the ecological impacts at individual sites, it is important to step back and consider a broader perspective. Because ground water recharge is an option typically pursued where water is scarce, planners must be aware that water would be similarly scarce for nonhuman components of the ecosystem. Thus, the decisions that are made about the water source for recharge and about how the project will be managed could cause other components of the ecosystem to face the consequences of reduced supplies and thus have widespread implications.

By adding to the storage and flow of ground water, artificial recharge modifies the hydrologic cycle. Water from the surface environment that otherwise would not have entered the ground water reservoir is emplaced underground through infiltration or injection techniques. This modification of the local water regime may have either beneficial or adverse consequences.

In areas where the base flow of streams is supported by ground water discharge, additions to the storage and flow of ground water by recharge may result in higher sustained streamflows during low flow or drought conditions. The flow of springs might also be sustained at higher levels through dry periods by the higher ground water heads that would result from artificial recharge. The increased ground water discharge to springs and streams would dampen somewhat the amplitude of their cyclical flow fluctuations, thus helping to sustain associated wetland environments. Wetlands not associated with surface water courses might also be sustained by the higher stand of the water table. In coastal areas the increased flow of streams and of direct ground water recharge to estuaries might help maintain less variable salinity conditions and to counter seasonal migration of the freshwater-saltwater interface in the stream channel and aquifer.

The extraction of ground water from some hydrogeologic settings can cause irreversible compaction of fine-grained beds of silt and clay in the aquifer, which in turn causes the land surface to subside. Through artificial recharge, ground water heads can be restored to or maintained at levels that can help prevent or reduce subsidence.

An indirect environmental impact may result from the fact that as ground water heads are raised by artificial recharge, less energy is used to pump a given quantity of water. This may result in a net savings in energy needs, depending on the recharging method employed and the energy required by the recharge operation. In addition, if the water used for recharge is of a higher quality than the ambient water in the aquifer, the quality of the recovered water may improve, resulting in a reduction in treatment requirements at the point of withdrawal. This, too, could result in some energy savings.

On the negative side, the diversion for ground water recharge of surface water or wastewater normally discharged to surface waters may result in a reduction of downstream flows, especially if the diversion is to another drainage basin or if the ground water reservoir does not discharge to the streams. Consequently, the reduced streamflow could result in undesirable changes to wetland environments and to salinity conditions in estuaries. Increased salinization of an estuary could result from the reduction in fresh streamflow and cause an undesirable change in the ecology of the estuary. San Francisco Bay and Florida Bay are examples of estuaries where ecological conditions have been altered by increases in salinity caused by reductions in fresh water inflows, although the diversions of fresh water in these areas are not attributable to ground water recharge facilities but are caused more broadly by increased demands for water. Stream pollution may increase and downstream appropriation rights may be jeopardized, as well, if recharge is not planned with an eye to the comprehensive needs of the region.

The raised level of the water table caused by artificial recharge sometimes can have deleterious consequences. If a water table is allowed to rise to the soil zone, soils may become water-logged and salinized and agricultural crops or native vegetation might be affected adversely. Underground structures, pipelines, gravel pits, and other facilities built during low stands of the water table could become inundated if the water table rises to their level. Dewatering schemes would be required to depress the water table locally and keep substructures dry should inundation or other adverse effects occur.

Perhaps the least predictable and the most difficult to remedy of all the potential environmental impacts of artificial recharge with wastewater is the likelihood of degrading ground water quality. Recharge with source waters of impaired quality could introduce microbial, inorganic, and organic chemical constituents into ground water, with the potential to cause environmental problems. As mentioned earlier, biochemical and geochemical reactions between the source water and the resident ground water and/or the aquifer materials could result in mobilization of chemical constituents that are part of the mineral framework of the aquifer. In addition, artificial recharge can leach anthropogenic contaminants from the vadose zone to ground water and move pollution plumes in aquifers to where they are not wanted. The resulting water quality degradation may require that the recovered ground water receive treatment not previously needed. Natural discharges to surface waters of ground water whose quality has been altered by wastewater recharge could also be damaging to the ecology of the receiving surface water body. The inability to identify all of the organic compounds in the recharge water, coupled with the difficulty of predicting the biochemical and geochemical changes in the subsurface, creates uncertainty with respect to the potential for degradation of ground water quality and the resulting environmental and ecological consequences.

SUMMARY

The growing competition for water in the United States will bring more and more attention to ground water resources. Artificial recharge of ground water is an established practice, and a long history of experimentation and actual full-scale recharge projects exists and clearly shows the benefits this technology can bring. Considerable experience-based, practical knowledge and experience about the problems and potentials of this technology is available (Asano, 1985). Thus, artificial recharge is likely to receive increased emphasis in the future, particularly regarding its role in the treatment and storage of municipal wastewater and other low-quality water for reuse. Given the broad body of knowledge available, the issues that arise with artificial recharge of ground water generally can be addressed through careful preproject planning and ongoing management.

There are still several operational issues that must be addressed on a site-specific basis. These concerns are related to project sustainability, treatment needs, public health impacts, and economic and institutional constraints. In the short-term, project sustainability is controlled by operating and managing the system so as to prevent or control clogging. Long-term sustainability is dependent on finding the best combination of pretreatment, soil-aquifer treatment, and posttreatment for determining whether the source waters will exceed the treatment and removal capacity of the soil-aquifer treatment system. These issues are discussed in the following chapters.

REFERENCES

Asano, T., ed. 1985. Artificial Recharge of Groundwater. Boston, Mass.: Butterworth. 767 pp.

Bouwer, H. 1978. Ground Water Hydrology. New York:McGraw-Hill. 480 pp.

Bouwer, H. 1982. Design considerations for earth linings for seepage control. Ground Water 20(5):531-537.

Bouwer, H., and R. C. Rice. 1984. Hydraulic properties of stony vadose zones. Ground Water 22(6):696-705.

Bouwer, H., and R. C. Rice. 1989. Effect of water depth in ground water recharge basins on infiltration rate. J. Irrig. Drain. Eng. 115(4):556-568.

Bouwer, H. 1990. Effect of water depth and ground water table on infiltration from recharge basins. Pp. 377-384 in: Proceedings of the 1990 National Conference of the Irrigation and Drainage Division. American Society of Civil Engineers, S. C. Harris ed. Durango, Colo. July 11-13, 1990.

Hultquist, R. H., R. H. Sakaji, and T. Asano. 1991. Proposed California regulations for ground water recharge with reclaimed municipal wastewater. Pp. 759-764 in: Proceedings of the 1991 Specialty Conference, Environmental Engineering, American Society for Civil Engineers. Reno, Nev. July 1991.

McCarty, P. L., B. E. Rittmann, and E. J. Bouwer. 1984. Microbiological processes affecting chemical transformations in ground water. Pp. 89-116 in Ground Water Pollution Microbiology, G. Bitton and C. P. Gerba eds. New York: John Wiley.

Peters, J. H., and C. Castell-Exner, eds. 1993. Proceedings Dutch-German Workshop on Artificial Recharge of Groundwater, Sept. 1993, Castricum, The Netherlands. Nieuwegein, The Netherlands: Keuringsinstituut voor Waterleidingartikelen.

2

Source Waters and Their Treatment

Water for human use comes from various sources—generally lakes, rivers, or underground aquifers. This report, which examines the potential of artificial recharge, uses the term "source water" to mean the recharge source—the water supplied to a surface infiltration or injection well recharge system. Potential source waters of impaired quality for artificial recharge include treated municipal wastewater, stormwater runoff, and irrigation return flows. The quality of source waters may be improved by the use of various pretreatment and disinfection processes. This chapter evaluates the quality of municipal wastewater and the quality improvement gained from primary, secondary, and advanced wastewater treatment. The quality of urban stormwater runoff and its possible treatment methods are also discussed, as is the quality of irrigation return flow and the problems inherent in treating it. Industrial wastewater and industrial stormwater runoff are not considered in depth; although industrial wastewater might at times be suitable for ground water recharge, its potential use would be extremely site specific and a general evaluation is not useful.

The quality of the source water considered for ground water recharge has a direct bearing on operational aspects of recharge facilities and also on the use to be made of the recovered water. The source water characteristics that affect the operational aspects of recharge facilities include suspended solids (SS), dissolved gases, nutrients, biochemical oxygen demand (BOD), microorganisms, and the sodium adsorption ratio (which affects soil permeability). The constituents that have the greatest possible adverse effects when the recharge is intended to support potable use include organic and metallic toxicants, nitrogen compounds, and pathogens.

MUNICIPAL WASTEWATER

Characteristics

The quantity and quality of wastewater delivered from varies among communities, depending on the number of commercial and industrial establishments in the area, the per capita in-house water use (which may vary from 400 l/day or more in industrialized countries to 40 l/day or less in developing or water-short countries), and the condition of the sewer system. Raw municipal wastewater may include contributions from domestic and industrial sources, infiltration and inflow from the collection system, and, in the case of combined sewer systems, urban stormwater runoff. The typical composition of untreated municipal wastewater appears in Table 2.1.

The occurrence and concentration of pathogenic microorganisms in raw wastewater depend on a number of factors, and it is not possible to predict with any degree of assurance what the general characteristics of a particular wastewater will be with respect to infectious agents. Important variables include the sources contributing to the wastewater, the original purpose of the water use, the general health of the contributing population, the existence of "disease carriers" in the population, and the ability of infectious agents to survive outside their hosts under a variety of environmental conditions. Table 2.2 lists infectious agents potentially present in untreated municipal wastewater. Table 2.3 illustrates the variety and order of magnitude of the concentration of microorganisms in untreated municipal wastewater.

Viruses are not normally excreted for prolonged periods by healthy individuals, and the occurrence of viruses in municipal wastewater fluctuates widely. Viral concentrations are generally highest during the summer and early autumn months. Viruses shed from an infected individual commonly range from 1,000 to 100,000 infective or plaque forming units (pfu's) per gram (g) of feces, but may be as high as 1,000,000 pfu/g of feces (Feachem et al., 1983). Viruses as a group are generally more resistant to environmental stresses than many of the bacteria, although some viruses persist for only a short time in municipal wastewater. Viral levels in the United States have been reported to be as high as 700 pfu/100 ml, but are typically less than 100 pfu/100 ml (American Society of Civil Engineers, 1970; Melnick et al., 1978).

Dissolved inorganic solids (total dissolved solids or salts, TDS) are not altered substantially in most wastewater treatment processes. In some cases, they may increase as a result of evaporation in lagoons or storage reservoirs. Therefore, unless wastewater treatment processes specifically intended to remove mineral constituents are employed, the composition of dissolved minerals in treated wastewater used for ground water recharge can be expected to be similar to the composition in the raw wastewater. The concentration of dissolved minerals in untreated wastewater is determined by the concentration in the domestic water

TABLE 2.1 Typical Contaminants in Untreated Municipal Wastewater

	Unit	Concentration		
		Weak	Medium	Strong
Solids, total	mg/l	350	720	1,200
Dissolved, total	mg/l	250	500	850
Fixed	mg/l	145	300	525
Volatile	mg/l	105	200	350
Suspended solids	mg/l	100	220	350
Fixed	mg/l	20	55	75
Volatile	mg/l	80	165	275
Setteable solids	mg/l	5	10	20
Biochemical oxygen demand[a]	mg/l	110	220	275
Total organic carbon (TOC)	mg/l	80	160	290
Chemical oxygen demand (COD)	mg/l	250	500	1,000
Nigrogen (total as N)	mg/l	20	40	85
organic	mg/l	8	15	45
Free ammonia	mg/l	12	25	50
Nitrites	mg/l	0	0	0
Nitrates	mg/l	0	0	0
Phosphorus (total as P)	mg/l	4	8	15
Organic	mg/l	1	3	5
Inorganic	mg/l	3	5	10
Chlorides[b]	mg/l	30	50	100
Sulfate[b]	mg/l	20	30	50
Alkalinity[c]	mg/l	50	100	NA
Grease	mg/l	50	100	150
Total coliform	no./100 ml	10^6–10^7	10^7–10^8	10^7–10^9
Volatile organic compounds	μg/l	< 100	100–400	> 400

Note: NA = not available.

[a] 5-day, 20°C (BOD_5, 20°C).
[b] Values should be increased by amount present in domestic water supply.
[c] As calcium carbonate ($CaCO_3$).

Source: Metcalf & Eddy, Inc., 1991.

supply plus mineral pickup resulting from domestic water use, which in the United States varies from 200 to 400 mg/l.

Wastewater treatment levels are generally classified as preliminary, primary, secondary, and advanced. The nature of each level of treatment is discussed in the following sections. (Note that the costs of treatment at least through secondary treatment is required before disposal to receiving waters and therefore would not be the responsibility of the recharge operation.)

TABLE 2.2 Infectious Agents Potentially Present in Untreated Municipal Wastewater

	Disease
Protozoa	
Entamoeba histolytica	Amebiasis (amebic dysentery)
Giardia lamblia	Giardiasis
Balantidium coli	Balantidiasis (dysentery)
Cryptosporidium	Cryptosporidiosis, diarrhea, fever
Helminths	
Ascaris lumbricoides (roundworm)	Ascariasis
Ancylostoma duodenale (hookworm)	Ancylostomiasis
Necator americanus (roundworm)	Necatoriasis
Ancylostoma (spp.) (hookworm)	Cutaneous larva migrans
Strongyloides stercoralis (threadworm)	Strongyloidiasis
Trichuris trichuria (whipworm)	Trichuriasis
Taenia (spp.) (tapeworm)	Teaniasis
Enterobius vermicularis (pinworm)	Enterobiasis
Echinococcus granulosus (spp.) (tapeworm)	Hydatidosis
Bacteria	
Shigella (4 spp.)	Shigellosis (dysentery)
Salmonella ryphi	Typhoid fever
Salmonella (1,700 serotypes)	Salmonellosis
Vibro cholerae	Cholera
Escherichia coli (enterophathogenic)	Gastroenteritis
Yersinia enterocolitica	Yersiniosis
Leptospira (spp.)	Leptospirosis
Legionella pneumopilla	Legionnaire's disease
Campylobacter jejune	Gastroenteritis
Viruses	
Enteroviruses (72 types) polio, echo, coxsackie, new enteroviruses	Gastroenteritis, heart anomalies meningitis, others
Hepatitis A virus	Infectious hepatitis
Adenovirus (47 types)	Respiratory disease, eye infections
Rotavirus (4 types)	Gastroenteritis
Parvovirus (3 types)	Gastroenteritis
Norwalk agent	Diarrhea, vomiting, fever
Reovirus (3 types)	Not clearly established
Astrovirus (5 types)	Gastroenteritis
Calicivirus (2 types)	Gastroenteritis
Coronavirus	Gastroenteritis

Source: Adapted from Sagik et al., 1978; Hurst et al., 1989.

TABLE 2.3 Microorganism Concentrations in
Untreated Municipal Wastewater

	Concentration (number per 100 ml)
Fecal Coliforms	$10^4 - 10^9$
Fecal streptococci	$10^4 - 10^6$
Shigella	$1 - 10^3$
Salmonella	$10^2 - 10^4$
Pseudomonas aeruginosa	$10^3 - 10^4$
Clostridium perfringens	$10^3 - 10^5$
Helminth ova	$1 - 10^3$
Giardia lamblia cysts	$10 - 10^4$
Cryptosporidium oocysts	$10 - 10^3$
Entamoeba histolytica cysts	$1 - 10^1$
Enteric viruses	$10^2 - 10^4$

Source: Adapted from various sources.

Primary Treatment

The first step in treatment, sometimes referred to as preliminary treatment, consists of the physical processes of screening, or comminution, and grit removal. Coarse screening is usually the first step and is used to remove large solids and trash that may interfere with later treatment processes. Comminution devices are sometimes used to cut up solids into a smaller size to improve downstream operations. Grit chambers are designed to remove material such as sand, gravel, cinders, eggshells, broken glass, seeds, coffee grounds, and large organic particles, such as food waste. Settling of most organic solids is prevented in the grit chamber because of the high flow velocity of wastewater through the chamber. Other preliminary treatment operations can include flocculation, odor control, chemical treatment, and pre-aeration.

Past this initial screening, primary treatment consists of physical processes to remove settleable organic and inorganic solids by sedimentation and floating materials by skimming. These also removes some of the organic nitrogen, organic phosphorus, and heavy metals. Additional phosphorus and heavy metal removal can be achieved through the addition of chemical coagulants and polymers. Primary treatment, together with preliminary treatment, typically removes 50 to 60 percent of the suspended solids and 30 to 40 percent of the organic matter. Primary treatment does not remove the soluble constituents of the wastewater.

Primary treatment has little effect on the removal of most biological species present in wastewater. However, some protozoa and parasite ova and cysts will settle out during primary treatment, and some particulate-associated microorganisms may be removed with settleable matter. Primary treatment does not reduce the level of viruses in municipal wastewater. While primary treatment by itself generally is not considered adequate for ground water recharge applications, primary effluent has been successfully used in soil-aquifer treatment systems at some spreading sites where the extracted water is to be used for nonpotable purposes (Carlson et al., 1982; Lance, Rice, and Gilbert, 1980; Rice and Bouwer, 1984). The higher organic content of primary effluent may enhance nitrogen removal by denitrification in the SAT system (Lance, Rice, and Gilbert, 1980) and may enhance removal of synthetic organic compounds by stimulating greater microbiological activity in the soil (McCarty, Rittman, and Bouwer, 1984). A disadvantage of using primary effluent is that infiltration basin hydraulic loading rates may be lower because of higher suspended solids and weaker biological activity on and in the soil of the infiltration system. Also, too much organic carbon in the recharge water can have adverse effects on processes that occur in the soil and aquifer systems. In most cases, wastewater receives at least secondary treatment and disinfection, and often tertiary treatment by filtration, prior to augmentation of nonpotable aquifers by surface spreading.

Secondary Treatment

Secondary treatment is intended to remove soluble and colloidal biodegradable organic matter and suspended solids (SS). In some cases, nitrogen and phosphorus also are removed. Treatment consists of an aerobic biological process whereby microorganisms oxidize organic matter in the wastewater. Several types of aerobic biological processes are used for secondary treatment, including activated sludge, trickling filters, rotating biological contactors (RBCs), and stabilization ponds. Generally, primary treatment precedes the biological process; however, some secondary processes are designed to operate without sedimentation, e.g., stabilization ponds and aerated lagoons. Typical microorganism removal efficiencies for activated sludge and trickling filter secondary treatment processes are given in Table 2.4. Concentration ranges for inorganic constituents and some other parameters in secondary-treated municipal wastewater are presented in Table 2.5. Information on the concentration of trace organics in activated sludge secondary effluent from the City of Phoenix's 23rd Avenue treatment plant is given in the "Phoenix, Arizona Projects" case study in Chapter 6.

The activated sludge process is considered to be a high-rate biological process because of the high concentrations of microorganisms used for the metabolization of organic matter. Trickling filters may be classified as either low-rate or high-rate based on their hydraulic and organic loading, mode of operation, and other factors. These processes accomplish biological oxidation in relatively small basins and use sedimentation tanks (secondary clarifiers) after

TABLE 2.4 Typical Percent Removal of Microorganisms by Conventional Treatment Processes

	Primary Treatment	Secondary Treatment	
		Activated Sludge	Trickling Filter
Fecal coliforms	< 10	0-99	85-99
Salmonella	0-15	70-99	85-99 +
Mycobacterium tuberculolosis	40-60	5-90	65-99
Shigella	15	80-90	85-99
Entamoeba histolytica	0-50	Limited	Limited
Helminth ova	50-98	Limited	60-75
Enteric viruses	Limited	75-99	0-85

Source: Crook, 1992.

the aerobic process to separate the microorganisms and other settleable solids from the treated wastewater.

In the activated sludge process, treatment is provided in an aeration tank in which the wastewater and microorganisms are in suspension and continually mixed through aeration. Trickling filters utilize media such as stones, plastic shapes, or wooden slats in which the microorganisms become attached. RBCs are similar to trickling filters in that the organisms are attached to support media, which in this case are partially submerged rotating discs in the wastewater stream.

These processes are capable of removing up to 95 percent of BOD, COD, and SS originally present in the wastewater and significant amounts of many (but not all) heavy metals and specific organic compounds (Water Pollution Control Federation, 1989). Trickling filters are not as effective as activated sludge processes in removing soluble organics because of less contact between the organic matter and microorganisms. Activated sludge treatment can reduce the soluble BOD fraction to 1 to 2 mg/l, while the trickling filter process typically reduces soluble BOD to 10 to 15 mg/l (U.S. Environmental Protection Agency, 1992). Biological treatment, including sedimentation, typically reduces the total BOD to 15 to 30 mg/l, COD to 40 to 70 mg/l, and TOC to 15 to 25 mg/l (U.S. Environmental Protection Agency, 1992). Very few dissolved minerals are removed during conventional secondary treatment.

Pond systems require relatively large land areas and are most widely used in rural areas and in warm climates and where land is available at reasonable cost. They are often arranged in a series of anaerobic, facultative, and maturation ponds with an overall hydraulic detention time of 10 to 50 days, depending on the design temperature and effluent quality required (Mara and Cairncross, 1989). Most organic matter removal occurs in the anaerobic and facultative ponds.

TABLE 2.5 Constituent Concentrations and other Parameters for Secondary-Treated Municipal Wastewater

	Concentration[a] (mg/l)
Calcium	9–84
Potassium	9–108
Magnesium	12–176
Sodium	44–1320
Amonium	0–501
Chlorine	43–2450
Fluoride	0.2–3.8
Bicarbonate	76–563
Nitrate	0.4–30
Phosphate	1.2–46
Sulfate	14–490
Silicondioxide	10–76
Hardness (as calcium carbonate)	62–951
pH (units)	6.3–8.4
Electrical conductivity	423-6570 µmho/cm
Total dissolved solids	210–4580
Arsenic	< 0.005–0.023
Boron	0.3–2.5
Cadmium	< 0.005–0.22
Chromium	<0.001-0.1
Copper	0.006–0.053
Lead	0.003–0.35
Molybdenum	0.001–0.018
Mercury	< 0.002–0.010
Nichol	0.003–0.60
Zinc	0.004–0.35
Biochemical oxygen demand	1.5-30
Chemical oxygen demand	40-70
Total suspended solids	10-25
Total organic carbon	15-25

[a]Concentration expressed in milligrams per liter unless otherwise noted.

Source: Treweek, 1985; Crook, 1992.

Maturation ponds, which are largely aerobic, are designed primarily to remove pathogenic microorganisms following biological oxidation processes. Well-designed stabilization pond systems are capable of reducing the BOD to 15 to 30 mg/l, COD to 90 to 135 mg/l, and SS to 15 to 40 mg/l (Shuval et al., 1986). The organic matter remaining in the effluent consists of soluble, biodegradable organic matter present in the raw wastewater, but not removed, plus intermediate products formed during the biological degradation of organic compounds and microbial cellular constituents (Metcalf & Eddy, Inc., 1991). The suspended solids are mainly organic in nature, consisting of biological solids produced during secondary treatment and other solids that escaped treatment and separation.

Stabilization ponds use algae to provide oxygen to the system. This system is considered a low-rate biological process. Mechanically aerated lagoon systems sometimes are used to provide secondary-level treatment. Stabilization ponds are capable of providing considerable nitrogen removal under certain conditions (e.g., high temperature and pH and long detention times) and are effective in removing microorganisms from wastewater. Well-designed and well-operated pond systems are capable of achieving a 6-log (99.9999 percent) reduction of bacteria, a 3-log (99.9 percent) reduction of helminths, and a 4-log (99.99 percent) reduction of viruses and cysts (Mara and Caimcross, 1989). Algae produced during pond treatment may present soil clogging problems during recharge.

Tertiary Treatment

The treatment of wastewater beyond the secondary or biological stage is sometimes called tertiary treatment. The term normally implies the removal of nutrients such as phos- phorus and nitrogen, and a high percentage of suspended solids. However, the term tertiary treatment is now being replaced in most cases by the term advanced wastewater treatment—which refers to any physical, chemical, or biological treatment process used to accomplish a degree of treatment greater than that achieved by secondary treatment.

Advanced Wastewater Treatment

Advanced wastewater treatment processes are designed to remove suspended solids and dissolved substances, either organic or inorganic in nature. Advanced wastewater treatment processes generally are used when a high-quality reclaimed water is necessary, such as for direct injection into potable aquifers. Commonly used processes and their principal removal functions are given in Table 2.6. The major advanced wastewater processes associated with ground water recharge are coagulation-sedimentation, filtration, nitrification, denitrification, phoshorus removal, carbon adsorption, and reverse osmosis.

Coagulation-Sedimentation

Chemical coagulation with lime, alum, or ferric chloride followed by sedimentation removes suspended solids, heavy metals, trace substances, phosphorus, and turbidity. Viral inactivation under alkaline pH conditions can be accom-

TABLE 2.6 Constituent Removal by Advanced Wastewater Treatment Operations and Processes

Principal Removal Function	Description of Operation or Process	Type of Wastewater Treated[a]
Suspended solids removal	Filtration	EPT, EST
	Microstrainers	EST
Ammonia oxidation	Biological nitrification	EPT, EBT, EST
Nitrogen removal	Biological nitrification/denitrification	EPT, EST
Nitrate removal	Separate-stage biological	EPT + nitrification denitrification
Biological phosphorus removal	Mainstream phosphorus removal[b]	RW, EPT
	Sidestream phosphorus removal	RAS
Combined nitrogen and phosphorus removal by biological methods	Biological nitrification/denitrification and phosphorus removal	RW, EPT
Nitrogen removal by physical or chemical methods	Air stripping	EST
	Breakpoint chlorination	EST + filtration
	Ion exchange	EST + filtration
Phosphorus removal by chemical addition	Chemical precipitation with metal salts	RW, EPT, EBT, EST
	Chemical precipitation with lime	RW, EPT, EBT, EST
Toxic compounds and refractory organics removal	Carbon adsorption	EST + filtration
	Powdered activated carbon	EPT
	Chemical oxiation	EST + filtration
Dissolved inorganic solids removal	Chemical precipitation	RW, EPT, EBT, EST
	Ion exchange	EST + filtration
	Ultrafiltration	EST + filtration
	Reverse osmosis	EST + filtration
	Electrodialysis	EST + filtration + carbon adsorption
Volatile organic compounds	Volatilization and gas stripping	RW, EPT

[a]EPT - effluent from primary treatment; EBT - effluent from bioloigcal treatment (before clarification); EST - effluent from secondary treatment; RW - raw (untreated sewage); and RAS - return activated sludge.

[b]Removal process occurs in the main flowstream as opposed to sidestream treatment.

Source: Metcalf & Eddy, Inc., 1991.

plished using lime as a coagulant, but pH values of 11 to 12 are required before significant inactivation is obtained.

Filtration

Filtration is a common treatment process used to remove particulate matter prior to disinfection. Filtration involves the passing of wastewater through a bed of granular media, which retain the solids. Typical media include sand, anthracite, and garnet. Removal efficiencies can be improved through the addition of certain polymers and coagulants. Table 2.7 presents average constituent removal efficiencies for filtration. The concentrations of organic and inorganic constituents at three water reclamation plants operated by the Sanitation Districts of Los Angeles County are shown in Tables 2.8 and 2.9, respectively. All three of the treatment plants have conventional activated sludge treatment followed by filtration and disinfection.

Treatment of biologically treated secondary effluent by chemical coagulation, sedimentation, and filtration has been demonstrated to remove more than 99 percent of seeded poliovirus (Sanitation Districts of Los Angeles County, 1977). This treatment chain reduces the turbidity of the wastewater to very low levels, thereby enhancing the efficiency of the subsequent disinfection process. Chemical coagulation and sedimentation alone can remove up to 99 percent of viruses, although the presence of organic matter can significantly decrease the amount of viruses removed. Direct filtration, that is, chemical coagulation and filtration without sedimentation, has also been shown to remove up to 99 percent of seeded poliovirus (Sanitation Districts of Los Angeles County, 1977). In one study, sand and dual-media filtration of secondary effluent, without coagulant addition prior to filtration, did not significantly reduce enteric viral levels (Noss et al., 1989). The primary purpose of the filtration step is not to remove viruses, but to remove protozoa and helminth eggs and floc and other suspended matter that may contain adsorbed or enmeshed microorganisms, thereby making the disinfection process more effective.

Chemical coagulation and filtration followed by chlorine disinfection to very low total coliform levels can remove or inactivate 5 logs (99.999 percent) of seeded poliovirus through these processes alone and subsequent to conventional biological secondary treatment can produce effluent essentially free of measurable levels of bacterial and viral pathogens (Sanitation Districts of Los Angeles County, 1977; Sheikh et al., 1990).

All parasitic cysts may not be removed by direct filtration. In one study, *Giardia* cysts were present in concentrations ranging from 3 to 7 cysts/100 liters after direct filtration (Casson et al., 1990). Rose and Carnahan (1992) found both *Giardia* and *Cryptosporidium* cysts and oocysts in reclaimed water after direct filtration of activated sludge effluent in about 20 percent of the samples collected. Filtered, secondary effluent from the Reedy Creek Wastewater Treat-

TABLE 2.7 Treatment Levels Achievable with Various Combinations of Advanced Wastewater Treatment Unit Processes

Treatment Process[a]	Typical Effluent Quality[b]						
	SS (mg/l)	BOD$_5$ (mg/l)	COD (mg/l)	Total N (mg/l)	NH$_3$-N (mg/l)	PO$_4$ as P (mg/l)	Turbidity (NTU)
AS+F	4-6	<5-10	30-70	15-35	15-25	4-10	0.3-5
AS+F+CA	<3	<1	5-15	15-30	15-25	4-10	0.3-3
AS/N (single stage)	10-25	5-15	20-45	20-30	1-5	6-10	5-15
AS/N-D (separate stages)	10-25	5-15	20-35	5-10	1-2	6-10	5-15
MS addition to AS	10-20	10-20	30-70	15-30	15-25	<2	5-10
MS addition to AS +N/D+F	<5-10	<5-10	20-30	3-5	1-2	<1	0.3-3
BP	10-20	5-15	20-35	15-25	5-10	<2	5-10
BNP+F	<10	<5	20-30	<5	<2	<1	0.3-3

[a] AS = activated sludge; F = granular-medium filtration; CA = carbon adsorption- N = nitrification; D = denitrification; MS = metal salt addition; BP = biological phosphorus removal; and BNP = biological nitrogen and phosphorus removal.

[b] SS = suspended solids; BOD5 = biochemical oxygen demand over 5 days; COD = chemical oxygen demand; N = nitrogen; NH3 = ammonia; P04 = phosphate; P = phosphorus; and NTU = Nephelometric Turbidity Units.

Source: Adapted from Metcalf & Eddy, Inc., 1991.

TABLE 2.8 Organic Constituent Concentrations in Oxidized, Filtered, Disinfected Reclaimed Water at Three Water Reclamation Plants

Constituent	Range of Concentration (µg/l)		
	Whittier Narrows	San Jose Creek	Pomona
Methylene chloride	3.3-5.3	2.1-3.6	ND-3.9
Chloroform	3.5-6.9	3.8-7.4	3.8-5.8
Bromodichloromethane	ND-0.8	ND-1.3	ND-2.0
Dibromochloromethane	ND	ND-0.6	ND-1.2
Bromoform	ND	ND	ND
Carbon tetrachloride	ND	ND	N-0.4
1,1-Dichloroethane	ND	ND	ND
1,2-Dichloroethane	ND	ND	ND
1,1,1-Dichloroethane	0.5 -2.3	ND-1.1	ND
1,1,2-Dichloroethane	ND	ND	ND
1,1-Dichloroethylene	ND	ND	ND
Trichloroethylene	ND	ND	ND
Tetrachloroethylene	ND-0.6	ND-33.6	ND-1.1
Benzene	ND	ND	ND
Toluene	ND-0.4	ND-1.2	ND
Chlorobenzene	ND	ND	ND
O-Dichlorobenzene	ND	ND-0.6	ND
M-Dichlorobenzene	ND	ND	ND
P-Dichlorobenzene	0.9-2.2	0.9-1.5	ND-1.1
Trans-1,2-dichloroethylene	ND	ND	ND
Bromomethane	ND	ND	ND
Chloroethane	ND	ND	ND
2-Chloroethylvinylether	ND	ND	ND
Chloromethane	ND	ND	ND
1,2-Dichloropropane	ND	ND	ND
Cis-1,3-dichloropropene	ND	ND	ND
Trans-1,3-dichloropropene	ND	ND	ND
1,1,2,2-Tetrachloropropene	ND	ND	ND
Vinyl chloride	ND	ND	ND
Bis(2-ethylhexyl)phthalate	ND-2	ND-1	ND-5
1,2,4-Trichlorobenzene	ND	ND	ND
2,4,6-Trichlorophenol	ND	ND	ND
2,4,5-Trichlorophenol	ND	ND	ND
2,3,4-Trichlorophenol	ND	ND	ND
2,3,6-Trichlorophenol	ND	ND	ND
3,4,5-Trichlorophenol	ND	ND-3.0	ND
Pentachlorophenol	ND	ND	ND
Phenanthrene	ND	ND	ND
Fluoranthrene	ND	ND	ND
Atrizine	ND	ND	ND
Simazine	ND	ND	ND
Phenylacetic acid	ND-3	ND-1	ND

continued on next page

TABLE 2.8 Continued

	Range of Concentration (μg/l)		
Constituent	Whittier Narrows	San Jose Creek	Pomona
Phenol	ND-1	ND-2	ND-1
DDT	ND	ND	ND
BHC	ND	ND	ND
Lindane	0.03-0.06	0.03-0.05	0.03-0 05
Heptachlor	ND-0.01	ND	ND
Heptachlor epoxide	ND	ND	ND
Aldrin	ND	ND	ND
Dieldrin	ND	ND	ND
Endrin	ND	ND	ND
Toxaphene	ND	ND	ND
PCB (Arochlor 1242)	ND	ND	ND
PCB (Arochlor 1254)	ND	ND	ND

Note: ND = none detected.

Source: Bookman-Edmonston Engineering, Inc., 1992.

TABLE 2.9 Physical Properties and Inorganic Constituent Concentrations in Oxidized, Filtered, Disinfected Reclaimed Water at Three Water Reclamation Plants

	Range of Concentration (mg/l unless otherwise noted)		
Constituent	Whittier Narrows	San Jose Creek	Pomona
Total dissolved solids	490-541	603-680	539-603
Electrical conductivity (μmbo/cm)	930-1450	1030-1380	1070-1130
Calcium	40-56	33-75	46-61
Magnesium	13-25	17-22	13-15
Sodium	96-112	120-146	116-132
Potassium	10-13	13-15	11-13
Bicarbonate	229-301	246-342	195-295
Sulfate	79-135	105-190	68-115
Chloride	67-94	131-160	92-144
Hardness	154-208	202-263	173-211
Alkalinity	188-247	202-280	160-242
Fluoride	0.73-1.01	0.40-0.63	0.25-0.45
Nitrate (as N)	1.3-3.4	1.95-5.01	1.5-3.9
Nitrite (as N)	0.15-1.11	0.52-1.11	0.43-1.85
Ammonia nitrogen (as N)	13.6-19.0	10.2-15.0	10.7-19.5
Organic nitrogen (as N)	1.4-2.7	1.0-2.2	1.3-3.4

TABLE 2.9 Continued

Constituent	Range of Concentration (mg/l unless otherwise noted)		
	Whittier Narrows	San Jose Creek	Pomona
Chemical oxygen demand	28-42	29-64	28-42
Biochemical oxygen demand	4-8	4-9	2-6
Total organic carbon	10-15	9-13	10-15
Iron	ND-0.07	0.03-0.09	0.04-0.07
Manganese	ND-0.01	0.02-0.03	ND-0.02
Arsenic	0.001-0.002	0.002-0.004	0.002-0.006
Barium	0.03-0.05	ND-0.06	0.03-0.04
Boron	0.27-0.60	0.16-0.68	0.09-0.57
Cadmium	ND-0.004	ND-0.004	ND-0.003
Chromium (hexavalent)	ND	ND-0.02	ND
Chromium (total)	ND	ND	ND
Copper	ND	ND	ND
Cyanide	ND-0.02	ND-0.03	ND
Lead	ND-0.001	ND-0.002	ND
Mercury	ND-0.0002	ND-0.0006	ND
Nickel	ND-0.04	ND-0.06	ND-0.04
Selenium	0.004-0.010	0.001-0.002	ND-0.004
Silver	ND	ND-0.005	ND
Zinc	0.07-0.13	0.05-0.06	0.02-0.04
pH	6.9-7.1	6.8-7.1	7.0-7.3
Color (CU)	17	15-20	25-46
Temperature (°F)	69-82	71-81	68-83
Turbidity (T = NTU)	1.1-2.1	0.8-2.1	1.1-2.0

Note: ND = none detected.

Source: Bookman-Edmonston Engineering, Inc., 1992.

ment Plant in Florida was found to contain both *Giardia* cysts and *Cryptosporidium* oocysts in 5 of 12 samples analyzed for the organisms. The positive samples had average parasite concentrations of 1 cyst or oocyst/100 liters (CH2M Hill, 1993). The viability of the parasitic cysts was not determined in any of these studies.

Several types of filtration systems are used at municipal wastewater treatment plants including conventional dual- or multi-media filters, mono-medium deep bed filters, automatic-backwashing shallow-bed travelling bridge filters, downflow contiuously backwashing filters, and upflow continuously backwashing filters. They are usually designed and operated to achieve low suspended solids or turbidity prior to disinfection. Most types of filters have been ddemonstrated to be capable of producing effluent meeting an average turbidity limit of 2 NTU (James M. Montgomery Consulting Engineers, Inc., 1979; Matsumoto and Tchobanoglous, 1981; Lang et al., 1986; and Weinschrott and Tchobanoglous, 1986).

Nitrification

Nitrification is the biological conversion of ammonia nitrogen sequentially to nitrite nitrogen and nitrate nitrogen. Nitrification does not remove significant amounts of nitrogen from the effluent; it merely converts it to another form. Nitrification can be done in many suspended and attached growth aerobic treatment processes when they are designed to foster the growth of nitrifying bacteria. In the traditional activated sludge process, it is accomplished by designing the process to provide a retention time for suspended solids that is long enough to prevent the slow-growing nitrifying bacteria from being washed out of the system. Nitrification also occurs in trickling filters that operate at low BOD/TKN (total kjeldahl nitrogen) ratios, either in combination with BOD removal or as a separate advanced process following any type of secondary treatment. A well-designed and well-operated nitrification process can produce an effluent containing 1 mg/l or less ammonia nitrogen. Ammonia nitrogen can also be removed from effluent by several chemical or physical treatment methods, such as air stripping, ion exchange, reverse osmosis, and breakpoint chlorination.

Denitrification

Denitrification removes nitrate nitrogen from the wastewater. As with ammonia removal, denitrification is usually best done biologically for most municipal applications. In biological denitrification, nitrate nitrogen is used by a variety of heterotrophic bacteria as the terminal electron acceptor in the absence of dissolved oxygen (anaerobic conditions). In the process, nitrate nitrogen is converted to nitrogen gas which escapes to the atmosphere. A carbonaceous food source is also required by the bacteria in these processes. Stoichiometrically, about 4 mg of organic carbon is required for every milligram of nitrate to be denitrified. Biological denitrification processes can achieve effluent nitrogen concentrations between 2 mg/l and 12 mg/l nitrate nitrogen. The effluent total nitrogen will be somewhat higher, depending on the concentrations of volatile suspended solids and soluble organic nitrogen present.

Phosphorus Removal

Phosphorus can be removed from wastewater by either chemical or biological methods, or a combination of the two. Chemical phosphorus removal is accomplished by precipitating the phosphorus from solution by the addition of iron, aluminum, or calcium salts. Biological phosphorus removal relies on the aerobic culturing of bacteria that will store excess amounts of phosphorus when exposed to anaerobic conditions in the treatment process. Chemical phosphorus removal can attain effluent orthophosphate concentrations of less than 0.1 mg/l, while biological phosphorus removal will usually produce an effluent phosphorus concentration between 1.0 and 2.0 mg/l.

Carbon Adsorption

One of the most effective advanced wastewater treatment processes for removing biodegradable and refractory organic constituents is the use of granular activated carbon (GAC). GAC can reduce the levels of synthetic organic chemicals in wastewater by 75 to 85 percent. The basic mechanism of removal is by adsorption of the organic compounds onto the carbon. Carbon adsorption preceded by conventional secondary treatment and filtration can produce an effluent with a BOD of 0.1 to 5.0 mg/l, COD of 3 to 25 mg/l, and TOC of 1 to 6 mg/l.

The major organic fraction adsorbed by activated carbon is in the low-molecular-weight range, generally less than 10,000. High-molecular-weight fractions, consisting of mostly polar organic compounds and compounds having molecular weights greater than 50,000, are adsorbed poorly. They can be displaced by more strongly adsorbed organics. Further, organic compounds having low solubilities adsorb better than those having high solubilities in the wastewater. Other molecules having highly branched structures are removed much more slowly than those of identical molecular weight, but with configurations that permit the coiling and compactness that result in high rates of diffusion into the pores of the carbon.

Granular activated carbon will remove several metal ions, particularly cadmium, hexavalent chromium, silver, and selenium. It has also been used to remove un-ionized species, such as arsenic and antimony, from an acidic stream, and it reduces mercury to low levels, particularly at low pH values.

Reverse Osmosis

Reverse osmosis (RO) is used mainly as a wastewater treatment process to remove suspended and dissolved solids (including microorganisms), either organic or inorganic. Removal is accomplished by the passage of wastewater through a semipermeable membrane. Constituent removal is influenced by the size, shape, chemical characteristics, and concentration of the chemical species as well as the physical and chemical characteristics of the feed water and type of RO unit employed. Because of the nature of the RO process, feed water must be of a fairly high quality (low suspended solids content) to prevent membrane clogging and deterioration.

In a homologous series of organic compounds, solute removals (i.e., rejection by the membranes) increases as the number of carbon atoms or molecular weight increases. The removal of organic compounds by RO membranes increases as the degree of molecule branching increases and as the cross-sectional area increases. Organic molecules having molecular weights greater than 200 are rejected by the membrane on the basis of size alone. Depending on membrane pore size, RO can also remove viruses and virtually all larger microorganisms.

The effectiveness of RO in removing inorganic constituents at Orange County Water District's Water Factory 21 is illustrated in Table 2.10. Reverse osmosis can readily remove more than 90 percent of the gross organics and with proper pretreatment can reduce TOC levels to less than 1 mg/l. Table 2.11 provides information on the removal of organic priority pollutants by both GAC and RO at Water Factory 21.

DISINFECTION

The most important process for the destruction of microorganisms is disinfection. Although, the most common disinfectant is chlorine, ozone (O_3) and ultraviolet (UV) radiation are other prominent disinfectants used at wastewater treatment plants. Other disinfectants, such as gamma radiation, bromine, iodine, and hydrogen peroxide, have been considered for the disinfection of wastewater but are not generally used because of economical, technical, operational, or disinfection efficiency considerations. Membrane processes (e.g., microfiltration, ultrafiltration, and reverse osmosis) have been shown to be effective in removing microorganisms, including viruses, from municipal wastewater, but again are not commonly used. The strategy in the selection and use of disinfectants for source waters prior to recharge should recognize the possibility that the nature and quantities of the disinfection by-products (DBPs) that may be formed are different from those in conventional water treatment. For example, both chlorine and ozone react in wastewater with organic precursors, which are likely to be greater in number and concentration than in freshwater sources of drinking water, to form DBPs. Accordingly, treatment of water for potable purposes is being modified to minimize the use of oxidizing disinfectants. However, in the treatment of water for nonpotable purposes, the numbers and concentrations of DBPs are of less concern because long-term ingestion is not an issue.

Disinfectants

Chlorine

The efficiency of disinfection with chlorine depends on the water temperature, pH, degree of mixing, time of contact, presence of interfering substances,

TABLE 2.10 Removal of Inorganic Constituents by Reverse Osmosis at Water Factory 21, Orange County, California

Constituent	Units	Concentration Before Reverse Osmosis	Concentration After Reverse Osmosis
Total dissolved solids	mg/l	1230	72
Sodium	mg/l	270	24
Potassium	mg/l	20	2.1
Magnesium	mg/l	4.3	< 0.1
Calcium	mg/l	98	1.9
Iron	µg/l	40	40
Manganese	µg/l	12	1.5
Silver	µg/l	0.1	50
Aluminum	µg/l	82	59
Barium	µg/l	23	1.5
Beryllium	µg/l	< 1.0	< 1.0
Cadmium	µg/l	2	< 1.0
Cobalt	µg/l	1.3	< 1.0
Chromium	µg/l	< 1.0	< 1.0
Copper	µg/l	1.3	< 1.0
Mercury	µg/l	< 0.5	< 0.5
Nickel	µg/l	19	1
Lead	µg/l	< 1.0	< 1.0
Selenium	µg/l	< 5.0	< 5.0
Zinc	µg/l	< 50	< 50
Ammonia (as N)	mg/l	23	2.6
Organic nitrogen	mg/l	< 0.1	< 0.1
Total kjeldahl nitrogen	mg/l	23	2.6
Total alkalinity	mg/l	20	13
Total hardness	mg/l	262	5
Fluoride	mg/l	0.5	0.16
Chloride	mg/l	373	40
Nitrate (as N)	mg/l	0.4	0.1
Sulfate	mg/l	431	4.8
Boron	mg/l	0.6	0.5
Silica	mg/l	14	< 1.0
Chemical oxygen demand	mg/l	27	1
Total organic carbon	mg/l	10.3	0.8
MBAS[a]	mg/l	0.24	0.06
Color	color units	13	< 3
Free chlorine	mg/l	< 0.1	< 0.1
Total chlorine	mg/l	< 0.3	< 0.3
Total coliform	CFU/100 ml	< 0.3	< 0.3
Fecal coliform	CFU/100 ml	< 1	< 1

[a]MBAS = methylene blue active substances, a measure of surfactant concentration.

Source: Wesner, 1991.

TABLE 2.11 Removal Organic Priority Pollutants at Water Factory 21, Orange County, California by Granular Activated Carbon (GAC) and Reverse Osmosis (RO)

	Secondary Effluent Concentration (μg/l)	Concentration Before GAC[a] (μg/l)	Concentration After GAC (μg/l)	Concentration After RO (μg/l)
Chloroform	3.5	7.9	10.2	1.8
Bromodichloromethane	0.46	3.9	4.4	0.7
Dibromochloromethane	0.71	2.1	2.8	0.7
Bromoform	0.46	0.9	0.8	0.3
1,1,1-Trichloroethane	4.8	0.17	0.18	0.05
Trichloroethylene	1.1	0.04	0.06	0.01
Tetrachloroethylene	3.6	0.14	0.71	0.06
Carbon tetrachloride	0.05	0.10	0.21	0.06
Chlorobenzene	0.13	0.13	0.07	0.07
1,3-Dichlorobenzene	0.25	0.03	0.01	0.01
1,4-Dichlorobenzene	1.9	0.14	0.04	0.02
1,2-Dichlorobenzene	0.74	0.09	0.05	0.03
1,2,4-Trichlorobenzene	0.31	0.06	0.01	0.06
Napthalene	0.11	0.13	0.03	0.06
Ethylbenzene	0.04	0.03	0.02	0.04
2,4-Dichlorophenol	0.16	0.01	<0.05	<0.05
2,4,6-Trichlorophenol	0.13	0.05	<0.05	<0.05
Pentachlorophenol	1.23	0.48	<0.05	<0.05
PCB (Arochlor 1242)	0.40	0.00	<0.05	<0.05
Lindane	0.11	0.06	<0.05	<0.05
DDT	0.01	0.05	<0.05	<0.05
Di-n-butyl phthalate	0.94	0.41	0.24	0.20
Diethyl phthalate	1.14	0.59	0.82	0.32
Bis(2-ethylhexyl)phthalate	11	1.2	0.63	1.2
Isophorone	0.30	0.05	<0.05	<0.05

Note: All values are geometric mean concentrations.

[a]Treatment before GAC includes biological secondary treatment, chemcial (lime) clarification, air stripping, chlorination, and filtration.

Source: McCarty et al., 1982.

concentration and form of chlorinating species, and the nature and concentration of the organisms to be destroyed. In general, bacteria are less resistant to chlorine than are viruses, which in turn are less resistant than parasite ova and cysts.

The chlorine dosage required to disinfect a wastewater to any desired level is greatly influenced by the constituents present in the wastewater. Some of the interfering substances are organic constituents, which consume the disinfectant; particulate matter, which protects microorganisms from the action of the disinfectant; and ammonia, which reacts with chlorine to form chloramines, a much less effective disinfectant species than free chlorine. In practice, the amount of chlorine added is determined empirically, based on desired residual and effluent quality. Chlorine, which in low concentrations is toxic to many aquatic organisms, is easily controlled in wastewater by dechlorination, typically with sulfur dioxide or thiosulfate. Chlorine has the disadvantage that it must be handled carefully, and the safety precautions required can be expensive.

Secondary effluent can be disinfected with chlorine to achieve very low levels of coliform bacteria, although complete destruction of pathogenic bacteria and viruses is unlikely to occur. Chlorination of secondary effluent that has received further treatment to remove suspended matter can produce wastewater that is essentially free of bacteria and viruses. Chlorine, at the normal concentrations used in wastewater treatment, may not destroy helminth eggs, *Giardia lamblia*, and *Cryptosporidium* species.

Ozone

Ozone is a powerful disinfecting agent and a powerful chemical oxidant in both inorganic and organic reactions. Due to the instability of ozone, it must be generated on site from air or oxygen carrier gas. Ozone destroys bacteria and viruses by means of rapid oxidation of the protein mass, and disinfection is achieved in a matter of minutes. Some disadvantages are that the use of ozone is relatively expensive and energy intensive, ozone systems are more complex to operate and maintain than chlorine systems, and ozone does not maintain a residual in water. Ozone is a highly effective disinfectant for advanced wastewater treatment plant effluent, and it removes color and contributes dissolved oxygen. It also breaks down recalcitrant organic compounds into more biodegradable compounds, which is advantageous for ground water recharge and soil-aquifer treatment.

Ultraviolet Radiation

Irradiation of wastewater with ultraviolet radiation for disinfection is potentially a desirable alternative to chemical disinfection, owing to its inactivating power for bacteria and viruses, affordable cost, and the absence of chemical disinfection by-products. Exposure of microorganisms to the appropriate amount of electromagnetic (EM) radiation disrupts the cells, genetic material and interferes with the reproduction process. Some bacteria have repair enzyme systems that are activated by similar EM energies, and thus disinfected waters may be repopulated by these particular bacteria after disinfection when exposed to light. UV disinfection for water and wastewater is the newest of the disinfection technologies and therefore valuable large scale field applications are still under study. However, the trend is toward more use of UV disinfection.

The effectiveness of UV radiation as a disinfectant where fecal coliform limits are on the order of 200 per 100 ml has been well established, as is evidenced by its use at more than 120 small to medium wastewater treatment plants in the United States (U.S. Environmental Protection Agency, 1986a). An increasing amount of research is being conducted on the ability of UV radiation to achieve high levels of disinfection. In one pilot study, a UV dose of 60 mWs/cm^2, or greater, consistently disinfected unfiltered secondary effluent to a total coliform level of 23 per 100 ml, or less, and a UV dose of at least 97 mWs/cm^2 consistently disinfected filtered secondary effluent to a total coliform level of 2.2 per 100 ml, or less (Snider et al., 1991). The study also indicated that filtration, which was effective in removing significant amounts of suspended solids and providing an effluent with a turbidity of less than 2 NTU, enhanced the performance of the UV disinfection. Because water after soil-aquifer treatment has a very low turbidity, UV radiation may be the method of choice in this case.

A recent study applied UV radiation (dose, 47 mWs/cm^2) in a full scale pilot plant to a blend of 70 percent secondary municipal wastewater effluent and 30 percent surface water with 90 to 99 percent reductions in total coliform organisms, *E. coli,* fecal streptococci, *Salmonella*, and coliphages (Dizer et al., 1993). The inactivating effect was counteracted by binding of the coliphage to suspended particles, prompting the authors to recommend that wastewater be clarified prior to UV disinfection. A similar pilot study investigated both secondary and tertiary effluent and the effects of UV on reclaimed water characteristics, microbial regrowth potential in transport lines, and photoreactivation of bacteria after UV treatment (Chen et al., 1993). At a radiation dose of 100 mWs/cm^2, total coliform concentrations were reduced below the 2.2 MPN/100 ml drinking water limit, and doses between 100 and 200 mWs/cm^2 had no significant effect on other wastewater characteristics, including TOC, soluble COD, total COD, and chlorine demand. No photoreactivation was observed in samples kept dark after UV exposure, and the extent of reactivation in samples exposed to the light decreased as UV dose increased. Heterotrophic plate count increases in the transport lines after UV irradiation were observed, although total coliforms were unchanged.

Studies of virus survival kinetics for UV doses up to 40 mWs/cm^2 conclude that UV irradiation can effectively inactivate viruses of public health concern in drinking water (Battigelli et al., 1993). Protozoan cysts (*Giardia, Cryptosporidium*) and bacterial spores were the most resistant to UV radiation, but the net public health risk, as noted in Chapter 4, is probably low because these organisms are large enough to be physically removed in most ground water systems. Other recent laboratory and pilot plant studies (CH2M Hill, 1992; Wilson, 1992; Dizer et al., 1993) indicate that UV radiation is very effective in inactivating enteric viruses and coliphages in water and tertiary-treated (i.e., filtered) wastewater. However, the effect of subsequent chlorination on dissolved organic carbon (DOC) surviving UV radiation is still unknown.

Disinfection By-Products

Disinfection by-products (DBP) are the chemical transformation products of the disinfection of water to remove pathogens. There are a number of uncertainties regarding disinfection by-product chemistry. These include uncertainties regarding the chemical nature of the reduced carbon macromolecules commonly present in global water systems, the reaction chemistry of these carbon molecules with various oxidant/disinfectants, the toxicological properties of the identified by-products and residual disinfectant chemicals, and the reactivity and stability of the by-products in the soil and underground environment. We do not have an equal understanding of the reaction by-product chemistry for the various disinfectants employed in water treatment. It is safe to say that chlorine has been the most widely used disinfectant, but that within the last 10 years, lower chlorine doses and alternative disinfectants such as monochloramine and ozone have become more popular because of the concerns over the toxicity of chlorine disinfection by-products (Bull and Kopfler, 1991). It is probably true that more is known of chlorine by-products (at least from traditional water sources) than is known for chloramine by-products and that chloramine is less reactive with native carbon. Some current research is focusing on ozone by-product identification and some new halogenated by-products (bromohydrins) are being identified (Cavanaugh et al., 1992 and Shukairy et. al., 1994) resulting presumably from ozone conversion of bromide to hypobromite.

Chlorine Disinfection

Chlorine gas reacts reversibly and rapidly with water-forming aqueous species (HOCl, OCl$^-$, Cl3$^-$, Cl$_2$O) containing the oxidizing power of the original chlorine molecule. These species can participate in virtually every major class of reaction with organic molecules, and in the natural water environment complex natural product organic molecules (dissolved organic matter, DOM) are commonly available, largely in the form of humic substances, which may account for as much as 40 to 50 percent of the dissolved organic carbon in terrestrial streams. As mentioned earlier, the coincidence in natural waters of appreciable concentrations of reduced carbon and chlorine species added from the disinfection process results in a variety of chlorinated and nonchlorinated reaction products (John et al., 1992 and Christman et al., 1983).

Trihalomethanes (THMs) may or may not be the most concentrated principal chlorination by-product, depending on pH, water source, and chlorine dose. A wide variety of other chlorination by-products have been identified, including haloacids, unhalogenated carboxylic acids, haloacetonitriles, halogenated aldehydes, ketones, and phenols, as well as small, but toxicologically interesting group of halogenated hydroxyfuranones. There is no guarantee that the most important by-products from a health effects standpoint have been identified, because the major, known by-products usually account for 10 percent or less of the total organic halogen at treatment plant dosage levels, and only about 50 percent at laboratory chlorination levels. Most researchers have reported that residues of chlorinated water samples are mutagenic in the Ames histidine reversion assay (Kronberg et al., 1993). Despite this general conclusion, only five mutagenic derivatives of a trichlorinated hydroxyfuranone (MX) have been identified unambiguously as chlorination by-products (MX, EMX, red-MX, ox-MX, ox-EMX), and at the concentration levels at which they are ordinarily produced, they account for only a minor fraction (less than 20 percent) of the overall mutagenicity of whole water residues. Research points to the generally inadequate characterization of the toxicological properties of DBPs and the fact that mutagenicity data, alone, are of little use in the quantitation of risk (Bull and Kopfler, 1991). It could be concluded that we have not identified the majority of mutagens in chlorinated drinking water, nor do we understand the overall importance of their human health risk. On the other hand, chlorine and other disinfectants have played an extremely important role in controlling waterborne infectious disease, and the hazards that could arise from the abolition of disinfection would far outweigh any benefits from reduced toxicological hazards (Bull and Kopfler, 1991).

Monochloramine Disinfection

It is generally believed that monochloramine is inherently less reactive with native organic carbon in water systems than is chlorine with similar DBP formation at lower concentration levels (Jensen et al., 1985). In water treatment industry practice, chloramine is formed via the sequential or simultaneous additions of chlorine and ammonia, and in cases where chlorine is added before ammonia, higher levels of DBPs are observed than for simultaneous addition, or for prior addition of ammonia. In any event, the level of DBP production is considerably less than for disinfection with chlorine. A laboratory study could not detect ether-extractable products following dosing a fulvic acid solution with monochloramine at pH 9 (Jensen et al., 1985). However, THM production is well documented from studies of actual water supplies and other traditional by-products have been measured (Amy et al., 1984). A unique DBP of toxicological significance has been reported, namely, cyanogen chloride, the concentration of which may be increased when ozone is used in conjunction with monochloramine (Krasner et al., 1989).

The similarity of the identified monochloramine DBPs to those from chlorine is probably due to the very slow hydrolysis of monochloramine in water. Monochloramine is not unreactive with organic molecules and has been shown to form nitriles and other products after reaction with aldehydes and ketones, to form chlorophenols from phenol when a reaction time of several days is permitted, and to participate in electrophilic addition to activated olefinic bonds. Although it is uncertain whether all DBPs from monochloramine have been detected, it is more certain that the major by-products have been identified (Jensen et al., 1985).

Ozone Disinfection

Ozone is reactive toward reduced organic carbon molecules in water and attacks organic molecules directly as molecular ozone, O_3, and via hydroxyl radicals produced from ozone decomposition. Molecular ozone is a more selective oxidant, attacking olefinic bonds (epoxide formation), phenols, and simple amines, whereas hydroxyl radicals will attack a wider variety of organic substrates, including aliphatic acids, ketones, and unactivated aromatic rings. High pH values and ultraviolet radiation promote ozone decomposition favoring the hydroxyl radical pathway, whereas low pH values and the presence of radical scavengers such as carbonate and bicarbonate ions promote the molecular ozone pathway. In water treatment practice, ozone is often used in conjunction with hydrogen peroxide (peroxone process) to increase the generation of hydroxyl radicals.

Ozone produces a variety of disinfection by-products that are generally less oxidized and less halogenated than chlorine by-products, including aldehydes, ketones, carboxylic acids, and unstable peroxides. The smaller molecular weight aldehydes are typically found in greater concentration (5 to 20 micrograms per liter ($\mu g/l$)) than the larger alkanals (i.e., C_6 to C_{10}, approximately 0.1 to 2 $\mu g/l$). There have been some reports of aromatic aldehydes, although these may be products from postchlorination (Glaze et al., 1993).

Ozone (like chlorine) oxidizes bromide-ion-producing hypobromous acid (HOBr). Although hypobromous acid is a weaker acid than hypochlorous acid (HOCl), and since the conjugate acid form of the hypohalite ions is a better nucleophile, HOBr reacts more rapidly with organic substrates than does HOCl. Thus, although ozone contains no halogen, brominated disinfection by-products are known, and, in general, brominated derivatives are more worrisome than chlorinated derivatives toxicologically. This circumstance may be fortuitous, however, because consumption of hypobromite at least prevents or reduces its conversion to bromate by residual ozone. Although ozone oxidizes OBr^- to BrO_3^- at a significant rate in dilute aqueous solution, this reaction may be inhibited if radical scavengers are present.

There has not been as much research effort in identifying bromination reaction by-products as there has been for chlorination by-products. It is known, however, that HOBr reacts readily with aquatic humic material, and the formation of bromoform, other brominated THMs, and bromoacetic acids, has been reported (Glaze, 1986 and Cavanaugh et al., 1992). More recently, other brominated DBPs such as bromopicrin, cyanogen bromide, and bromoacetones, have been detected. A new group of labile brominated organic DBPs, the bromohydrins have been reported (Cavanaugh et al., 1992). The most abundant brominated alcohol has been assigned the structure of 3-bromo-2-methyl-2-butanol.

There is no direct toxicological concern for ozone because it is not expected to persist to the point of drinking water consumption.

Chlorine Dioxide Disinfection

Most of the reaction by-products of chlorine dioxide with aquatic humic materials appear to be monobasic and dibasic aliphatic acids, although a variety of polybasic aromatic acids have also been reported. Relatively few chlorinated products have been identified other than monochloromalonic, monochlorosuccinic, and dichloroacetic acids, all of which are observed with chlorination (Colclough et al., 1983). Both chlorite and chlorate have been shown to be present in waters disinfected with chlorine dioxide, and owing to the toxicological significance of chlorite (oxidative hemolytic anemia), measures should be taken to remove both chlorine dioxide and chlorite from finished waters prior to ingestion. The relative difficulty in preparing chlorine dioxide, the need for chlorite removal and the unknown consequences of dosing waters with the required reducing agents (themselves of unknown toxicological concern) would appear to obviate consideration of chlorine dioxide for disinfection in ground water recharge programs.

URBAN STORMWATER RUNOFF

Characteristics

Urban stormwater runoff can be a candidate for ground water recharge because of its close proximity to points of use and water supply infrastructure and because substantial water volumes are associated with urban runoff. Unfortunately, some stormwaters from urban areas may be badly polluted, requiring either careful selection of the water to be used for recharge or significant treatment before recharge, especially if the water is to be used for portable purpose. Stormwater runoff is also erratic in timing and quantity. Many studies have investigated urban stormwater runoff quality, with the Environmental Protection Agency's (EPA) Nationwide Urban Runoff Program (NURP) providing the largest and best-known database (U.S. Environmental Protection Agency, 1983). Unfortunately, the extensive analytical results reported by NURP and other studies have not included many of the pollutants that are likely to cause the greatest concern in ground water contamination or recharge studies.

Urban runoff comprises many different flow types. These may include dry weather base flows, urban stormwater runoff, combined sewer overflows (CSOs), and snowmelt. The relative magnitudes of these discharges vary considerably based on a number of factors. Season (especially cold versus warm weather) and land use have been identified as important factors affecting base flow and stormwater runoff quality, respectively (Pitt and McLean, 1986). This section summarizes a number of observations of runoff quality for these different flow types and land uses, along with observations of source area flows contributing to these combined discharges. This information can be used to identify the best urban stormwater runoff candidates for ground water recharge and the ones to avoid.

Land development increases urban stormwater pollutant concentrations and runoff water volumes. Impervious surfaces, such as rooftops, driveways, and roads, reduce infiltration of rainfall and runoff into the ground and degrade runoff quality. The average runoff volume from subdivisions has been reported to be more than 10 times greater than that of typical pre-development agricultural areas (Madison et al., 1979).

Hydraulic factors affecting runoff water volume (and therefore the amount of water available for recharge) include rainfall quantity and intensity, slope, soil permeability, land cover, impervious area, and depression storage. Research during NURP showed that the most important hydraulic factors affecting urban runoff volume were the quantity of rain and the extent of impervious surfaces directly connected to a stream or drainage system (U.S. Environmental Protection Agency, 1983). Directly connected impervious areas include paved streets, driveways, and parking areas draining to curb and gutter drainage systems, or roofs draining directly to a storm sewer pipe.

Table 2.12 presents a summary of the NURP stormwater data collected from about 1979 through 1982 (U.S. Environmental Protection Agency, 1983). BOD and nutrient concentrations in urban stormwater are relatively close in quality to those of typically treated municipal wastewater. However, as shown later, urban stormwater has relatively high concentrations of bacteria, along with high concentrations of many metallic and some organic toxicants. Land use and source areas (parking areas, rooftops, streets, landscaped areas, and so on) can all have important effects on urban stormwater runoff quality.

Bacterial Characteristics

Most descriptions of bacterial characteristics of urban runoff focus on fecal coliform analysis because of its historical use in water quality standards. Fecal coliform bacterial observations have long been used as an indicator of municipal wastewater contamination and therefore as an indicator of possible pathogenic microorganism contamination (Field and O'Shea, 1993). Fecal streptococcal analyses are also relatively common for urban runoff. Unfortunately, relatively few analyses of specific pathogenic microorganisms have been made for urban runoff.

Pathogenic bacteria have been found in urban runoff at many locations and are probably from several different sources (e.g., Field et al., 1976; Olivieri et al., 1977; Qureshi and Dutka, 1979; Environment Canada, 1980; Pitt, 1983; Pitt

TABLE 2.12 Median Stormwater Pollutant Concentrations for All Sites by Land Use (Nationwide Urban Runoff Program, NURP)

	Residential		Mixed Land Use		Commercial		Open/Nonurban	
	Median	COV	Median	COV	Median	COV	Median	COV
Biochemical oxygen demand (mg/l)	10	0.41	7.8	0.52	9.3	0.31	-	-
Chemical oxygen demand (mg/l)	73	0.55	65	0.58	57	0.39	40	0.78
Total suspended solids (mg/l)	101	0.96	67	1.14	69	0.85	70	2.92
Total Kjeldahl nitrogen (µg/l)	1900	0.73	1288	0.50	1179	0.43	965	1.00
Nitrite nitrogen plus nitrate nitrogen (µg/l)	736	0.83	558	0.67	572	0.48	543	0.91
Total phosphorus (µg/l)	383	0.69	263	0.75	201	0.67	121	1.66
Soluble phosphorus (µg/l)	143	0.46	55	0.75	80	0.71	26	2.11
Total Lead (µg/l)	144	0.75	114	1.35	104	0.68	30	1.52
Total Copper (µg/l)	33	0.99	27	1.32	29	0.81	-	-
Total Zinc (µg/l)	135	0.84	154	0.78	226	1.07	195	0.66

Note: COV = coefficient of variation - standard deviation/mean.

Source: U.S. Environmental Protection Agency, 1983.

and McLean, 1986; Field and O'Shea, 1993). Table 2.13 summarizes the occurrence of various pathogenic bacterial types found in urban stormwater at Burlington, Ontario; Milwaukee; Baltimore; and Cincinnati. The observed ranges of concentrations and percentage isolations of these biotypes vary significantly from site to site and at the same location for different times. However, many potentially pathogenic bacterial types can be present in urban stormwater runoff. As an example, the occurrence of *Salmonella* biotypes generally is low, and their reported density is usually less than one organism per 100 ml while *Pseudomonas aeruginosa* organisms are frequently encountered at densities greater than one thousand organisms per 100 ml (Pitt and McLean, 1986).

Salmonella has been reported in some, but not all, urban stormwaters (Qureshi and Dutka, 1979; Olivieri et al., 1977). Typical concentrations were from 5 to 300 Salmonella organisms per 10 liters. The types of *Salmonella* found in southern Ontario were *S. thompson* and *S. tvphimurium* var. *copenhagen* (Qureshi and Dutka, 1979). Almost all of the urban stormwater samples that had fecal coliform concentrations greater that 2,000 organisms per 100 ml had detectable *Salmonella* concentrations. However, about 27 percent of the samples having fecal coliform concentrations less than 200 organisms per 100 ml had detectable *Salmonella*. Other research, however, did not find significant correlations of *Salmonella* isolations with fecal coliform concentrations (Schillinger and Stuart, 1978).

Evidence has been found that *Shigella* is present in urban runoff and receiving waters and could present a significant health hazard (Olivieri et al., 1977). *Shigella* species causing bacillary dysentery are one of the primary human enteric disease producing bacterial agents present in water. The infective dose of *Shigella* necessary to cause dysentery may be quite low (it can be as low as 10 to 100 organisms, although the median infective dose is 10,000 and the infective dose will of course vary depending on the host's age, health, and other factors) (Feachem et al., 1983). Because of this low infective dose and the assumed presence of *Shigella* in urban waters, it may be a significant health hazard associated with urban runoff.

The most abundant of the potentially pathogenic bacteria that can be found in urban runoff and streams is *Pseudomonas aeruginosa* (Olivieri et al., 1977). This opportunistic pathogen is widely found in aquatic environments; it is associated with eye and ear infections and is resistant to antibiotics. This biotype has been detected frequently in urban runoff studies in concentrations that may cause potential infections. Typical populations of 1,000 to 10,000 organisms per 100 mL have been frequently reported in urban stormwater (Pitt and McLean, 1986). *E. coli* also is common in urban stormwater. Viruses may also be important pathogens in urban runoff, although they are usually present at low levels (Olivieri et al., 1977).

TABLE 2.13 Microorganisms Found in Urban Stormwater (organisms per 100 ml)

City, Province/State	Catchment/ Land-use	Staphylococcus aureus	Pseudomonas aeruginosa	Salmonella	Streptococcus species	Enterovirus (PFU/10 liters)	Others	Reference
Burlington, Ontario	Aldershot Plaza		14–3,000	S. seftenberg & S. newport isolated			Total fungi: 2×10^4–2×10^6	Qureshi and Dutka, 1979
	Malvern Road		1–740	100% negative			Total fungi: 9–400 Heterotroph count: 4×10^5–2×10^7	Qureshi and Dutka, 1979
Milwaukee, Wisconsin	Highway runoff	all < 1,000	all < 1,000	45% positive				Gupta, et al. 1981
Baltimore, Maryland	Bush St.	1,200	20	0.03	560,000 (FS)	6.9		Olivieri, et al., 1977
	Northwood	120	6	0.006	50,000 (FS)	170		
Cincinnati, Ohio	Business district				79% positive[a]			Geldreich and Kenner, 1969
	Residential area				80% positive[b]			Geldreich and Kenner, 1969
	Rural area				87% positive[c]			Geldreich and Kenner, 1969

[a]Streptrococcal bacteria types found: S. bovis/S. equinus (2%); Atypical S. faecalis (1%); S. faecalis liquifaciens (18%); and S. thompson: 4,500/100 ml.
[b]Streptococcal bacteria types found: S. bovis/S. equinus (0.5%); Atypical S. faecalis (1%); and S. faecalis liquifaciens (18%).
[c]Streptococcal bacteria types found: S. bocis/S. equinus (0.5%); Atypical S. faecalis (0.2%); and S. faecalis liquifaciens (12%).

Important Toxicants

Urban stormwater research has quantified some inorganic and organic hazardous and toxic substances frequently found in urban runoff (U.S. Environmental Protection Agency, 1983; Pitt and McLean, 1986). The NURP data (Table 2.14), collected from mostly residential areas throughout the United States, did not indicate any significant regional differences in the substances detected or in their concentrations (U.S. Environmental Protection Agency, 1983). However, residential and industrial data from Toronto showed significant concentration and yield differences for these two distinct land uses and for dry weather and wet weather urban runoff flows (Pitt and McLean, 1986).

The concentrations of many of these toxic pollutants exceeded the EPA ambient water quality criteria for human health protection by large amounts. As an example, typical standards for polycyclic aromatic hydrocarbons (PAHs) in surface waters used as drinking water supplies are 2.8 nanograms per liter (ng/l) (U.S. Environmental Protection Agency, 1986b). As shown on Table 2.14, urban runoff concentrations of chrysene (600 to 10,000 ng/l), fluoranthene (300 to 21,000 ng/l), phenanthrene (300 to 10,000 ng/l) and pyrene (300 to 16,000 ng/l) (four of the most common PAHs found in urban runoff) have been reported to be from 100 to almost 10,000 times greater than this criterion. Even though most of the PAHs are associated with particulate solids, filterable concentrations of these PAHs are still likely to be many times greater than this criterion.

Table 2.15 lists toxic and hazardous substances that are generally found in more than 10 percent of industrial and residential urban runoff samples analyzed (Galvin and Moore, 1982; U.S. Environmental Protection Agency, 1983; Pitt and McLean, 1986). Available NURP data do not reveal that toxic urban runoff conditions differ significantly in parts of the United States (U.S. Environmental Protection Agency, 1983). The pesticides shown were found mostly in urban runoff from residential areas, while heavy metals and other hazardous materials were much more prevalent in industrial areas. Urban runoff dry weather base flows may also be contributors of hazardous and toxic pollutants. Lindane and dieldrin may be very important in residential dry weather flows, while polychlorinated biphenyls (PCBs) may be very important in industrial dry weather flows. Many of the heavy metals found in industrial urban runoff were high during both dry weather and wet weather conditions.

Contamination by Industrial Wastewater

The potential for toxicants and hazardous materials in industrial areas to contaminate urban stormwater runoff is a serious problem. Inappropriate discharges of industrial wastewaters to storm drainage systems are relatively common and can cause serious contamination. Usually, the wastewater streams from all steps in the manufacturing process are collected in one location prior to

TABLE 2.14 Summary of National Urban Runoff Program Priority Pollutant Analyses

	Frequency of Detection (%)	Range of Detected Concentrations (μg/l)
Pesticide		
α-BHC	20	0.0027 to 0.1
y-BHC (lindane)	15	0.007 to 0.1
Chlordane	17	0.01 to 10
α-Endosulfan	19	0.008 to 0.2
Metals and Inorganics		
Antimony	13	2.6 to 23
Arsenic	52	1 to 51
Beryllium	12	1 to 49
Cadmium	48	0.1 to 14
Chromium	58	1 to 190
Copper	91	1 to 100
Cyanides	23	2 to 300
Lead	94	6 to 460
Mercury	10	0.6 to 1.2
Nickel	43	1 to 182
Selenium	11	2 to 77
Zinc	94	10 to 2400
PCBs and Related Compounds (detected in less than 1% of all samples)		
Halogenated Aliphatics		
Methylene chloride	11	5 to 15
Ethers (none detected in any of the samples)		
Monocyclic Aromatics (detected in less than 6% of all samples)		
Phenols and Cresols		
Phenol	14	1 to 13
Pentachlorophenol	19	1 to 115
4-Nitro phenol	10	1 to 37
Phthalate Esters		
bis(2-ethylhexyl) phthalate	22	4 to 62
Polycyclic Aromatic Hydrocarbons		
Chrysene	10	0.6 to 10
Fluoranthene	16	0.3 to 21
Phenanthrene	12	0.3 to 10
Pyrene	15	0.3 to 16

Note: Analyses are based on 121 samples from 17 cities. Only those compounds found in greater than 10% of outfall samples are shown.

Source: U.S. Environmental Protection Agency, 1983.

TABLE 2.15 Hazardous and Toxic Substances Found in Urban Runoff

	Residential Areas	Industrial Areas
Halogenated Aliphatics		
1,2,-Dichloroethene		x
Methylene chloride		x
Tetrachloroethylene		x
Phthalate		
Bis(2-ethylene)phthalate	x	
Butylbenzyl phthalate	x	x
Diethyl phthalate		x
Di-N-butyl phthalate	x	x
Polycyclic Aromatic Hydrocarbons		
Phenanthrene		x
Pyrene		x
Other Volatiles		
Benzene	x	x
Chloroform		x
Ethylbenzene		x
N-Nitro-sodimethylamine		x
Toluene		x
Heavy Metals		
Aluminum	x	x
Chromium		x
Copper	x	x
Lead	x	x
Zinc	x	x
Pesticides and Phenols		
BHC	x	
Chlordane	x	
Dieldrin	x	
Endosulfan sulfate	x	
Endrin	x	
Isophorone	x	
Methoxychlor	x	
PCB-arochlor 1254		x
PCB-arochlor 1260		x
Pentachlorophenol	x	x
Phenol	x	x

Note: Substances were found in at least 10 percent of the stormwater samples analyzed.

Sources: Galvin and Moore, 1982; U.S. Environmental Protection Agency, 1983; Pitt and McLean, 1986.

treatment and disposal. These wastewaters originate from many different areas of the plant and/or steps of the industrial process. Activities and areas of a plant that are likely to discharge contaminated wastewater into the storm drainage system include loading and unloading operations, outdoor storage or processing, cooling or process wastewater discharges, particle-generating processes, and illicit or inadvertent connections to the storm drainage system (Pitt et al., 1993). Industrial runoff usually is much more polluted by toxicants than runoff from other land uses. Many of these compounds may present serious problems in the operation of a ground water recharge facility and in ensuring the safety of the water for later reuse. In addition, the behavior of many of these compounds in the soil-land-aquifer system is not well known. Industrial stormwater should therefore not be considered an appropriate source water for ground water recharge.

Relative Contributions of Different Flow Periods

Tables 2.16 and 2.17 summarize residential/commercial and industrial urban runoff characteristics during both warm and cold weather in Toronto (Pitt and McLean, 1986). These tables show the relative importance of wet weather and dry weather flows coming from separate urban stormwater systems. Possibly 25 percent of all separate urban stormwater outfalls have water flowing in them during dry weather, and as many as 10 percent of them are grossly contaminated with raw municipal wastewater and industrial wastewaters (Pitt et al., 1993). EPA's Stormwater Permit program requires municipalities to conduct urban stormwater outfall surveys to identify, and then correct, inappropriate discharges into separate storm drainage. However, substantial outfall contamination probably will exist for many years. If urban stormwater is recharged before it enters the drainage system (such as by using French drains, infiltration trenches, grass swales, porous pavements or percolation ponds in upland areas), the effects of contamination problems in the drainage system on ground water recharge will be reduced substantially. If outfall waters are to be recharged in larger regional facilities, then these periods of dry weather flows will have to be considered.

Similar problems occur in areas having substantial snowfalls. Table 2.18 (Part A and B) summarizes Toronto snowmelt and cold weather base flow characteristics (Pitt and McLean, 1986). The bacteria densities during cold weather are substantially less than during warm weather, but are still relatively high (U.S. Environmental Protection Agency, 1983). However, chloride and dissolved solids concentrations are much higher during cold weather. Early spring urban stormwater events also contain high dissolved solids concentrations (R. Bannerman, personal communication, 1993). Upland infiltration devices do not work well during cold weather because of freezing soils. Outfall flows occur under ice into receiving waters (including detention ponds) and may enter regional ground water recharge devices if not specifically diverted.

TABLE 2.16 Median Concentrations Observed at Toronto Outfalls During Warm Weather

	Warm Weather Baseflow		Warm Weather Stormwater	
	Residential	Industrial	Residential	Industrial
Stormwater volume (m^3/ha/season)			950	1,500
Base flow volume (m^3/ha/season)	1,700	2,100		
Total residue	979	554	256	371
Filterable residue (TDS)	973	454	230	208
Particulate residue (SS)	< 5	43	22	117
Chlorides	281	78	34	17
Total phosphorus	0.09	0.73	0.28	0.75
Phosphates	< 0.06	0.12	0.02	0.16
Total Kjeldahl nitrogen (organic nitrogen plus ammonia)	0.9	2.4	2.5	2.0
Ammonia nitrogen	< 0.1	< 0.1	< 0.1	< 0.1
Chemical oxygen demand (COD)	22	108	55	106
Fecal coliform bacteria (no./100 ml)	33,000	7,000	40,000	49,000
Fecal streptoccal bacteria (no./100 ml)	2,300	8,800	20,000	39,000
Pseudomonas aeruginosa bacteria (#/100 ml)	2,900	2,380	2,700	11,000
Arsenic	< 0.03	< 0.03	< 0.03	< 0.03
Cadmium	< 0.01	< 0.01	< 0.01	< 0.01
Chromium	< 0.06	0.42	<0.06	0.32
Copper	0.02	0.05	0.03	0.06
Lead	< 0.04	< 0.04	< 0.06	0.08
Selenium	< 0.03	< 0.03	< 0.03	< 0.03
Zinc	0.04	0.18	0.06	0.19
Phenolics (µg/l)	< 1.5	2.0	1.2	5.1
α-BHC (ng/l)	17	< 1	1	3.5
γ-BHC (lindane) (ng/l)	5	< 2	< 1	< 1
Chlordane (ng/l)	4	< 2	< 2	< 2
Dieldrin (ng/l)	4	< 5	< 2	< 2
Pentachlorobiphenol (PCB) (ng/l)	< 20	< 20	< 20	33
Pentachlorophenol (PCP) (ng/l)	280	50	70	705

Note: Concentrations are given in milligrams per liter unless otherwise indicated.

Source: Pitt and McLean, 1986.

Pollutant Contributions from Different Urban Source Areas

Sheet flow quality data are available from studies conducted in California, Washington, Nevada, Wisconsin, Illinois, Ontario, Colorado, New Hampshire, New York, and Alabama since 1979. A relatively large amount of parking and roof runoff quality data has been obtained from all of these locations, but only a

TABLE 2.17 Median Concentrations Observed at Toronto Outfalls During Cold Weather

	Cold Weather Baseflow		Cold Weather Snowmelt	
	Residential	Industrial	Residential	Industrial
Stormwater volume (m³/ha/season)			1,800	830
Base flow volume (m³/ha/season)	1,100	660		
Total residue	2,230	1,080	1,580	1,340
Filterable residue (TDS)	2,210	1,020	1,530	1,240
Particulate residue (SS)	21	50	30	95
Chlorides	1,080	470	660	620
Total phosphorus	0.18	0.34	0.23	0.50
Phosphates	< 0.05	< 0.02	< 0.06	0.14
Total Kjeldahl nitrogen (organic nitrogen plus ammonia)	1.4	2.0	1.7	2.5
Ammonia nitrogen	< 0.1	< 0.1	0.2	0.4
Chemical oxygen demand (COD)	48	68	40	94
Fecal coliform bacteria (no./100 ml)	9,800	400	2,320	300
Fecal streptoccal bacteria (no./100 ml)	1,400	2,400	1,900	2,500
Pseudoomonas aeruginosa bacteria (#/100 ml)	85	55	20	30
Cadmium	< 0.01	< 0.01	< 0.01	0 01
Chromium	< 0.01	0.24	< 0.01	0.35
Copper	0.02	0.04	0.04	0.07
Lead	< 0.06	< 0.04	0.09	0.08
Zinc	0.07	0.15	0.12	0.31
Phenolics (µg/l)	2.0	7.3	2.5	15
α-BHC (ng/l)	NA	3	4	5
γ-BHC (lindane) (ng/l)	NA	NA	2	
Chlordane (ng/l)	NA	NA	11	2
Dieldrin (ng/l)	NA	NA	2	NA
Pentachlorobiphenol (PCB) (ng/l)	NA	NA	NA	40

Note: Concentrations are given in milligrams per liter unless otherwise indicated. N.A. = not analyzed.

Source: Pitt and McLean, 1986.

few of these studies evaluated a broad range of source areas or land uses. However, there is adequate information to identify which upland areas are preferable for use as sources for ground water recharge and which ones to avoid because of excessive contamination. The major urban source area categories that have been studied include the following:

- roofs,
- paved parking areas,

TABLE 2.18 (Part A) Toronto Cold Weather Snowmelt Source Area Sheet flow Quality (median observed concentrations, mg/l)

Source Area	Total Solids	Filterable Solids	Suspended Solids	Reactive Chlorides	Total Phosphorus	Phosphates	Total Kjeldahl Nitrogen	Ammonia	Chemical Oxygen Demand
Industrial									
Pervious areas									
Grass/open areas	390	282	77	100	0.33	0.10	1.4	<0.1	47
Unpaved storage/ parking	2,925	1000	2105	113	1.1	0.46	5.3	0.2	160
Impervious areas									
Sidewalks	1,050	200	847	48	0.45	0.20	1.6	<0.1	63
Paved parking, storage, etc.	1,690	349	392	260	0.55	0.18	3.8	<0.1	135
Road gutters	1,320	575	625	230	0.60	0.15	1.8	<0.1	230
Residential/Commercial									
Pervious areas									
Grass/open areas	94	78	40	4.0	0.29	0.20	1.2	0.4	26
Impervious areas									
Sidewalks	390	29	281	6.4	0.63	0.38	2.6	2.6	98
Paved parking, driveways, etc.	918	274	380	81	0.64	0.08	2.5	<0.1	110
Paved roads	890	166	284	56	0.30	0.06	1.8	<0.1	140
Road gutters	530	190	152	25	0.54	0.28	2.3	<0.1	66
Roadside grass swales	380	155	50	37	0.59	0.17	1.8	0.1	40

TABLE 2.18 (Part B) Toronto Cold Weather Snowmelt Source Area Sheet Flow Quality (median observed concentrations, mg/l)

Source Area	Fecal Coliforms	Fecal streptococca[a]	Pseudomonas aeruginosa	Cadmium	Chromium	Copper	Lead	Zinc	Phenolics µg/l
Industrial									
Pervious Areas:									
Grass/open areas	< 20	100	< 20	< 0.005	0.01	0.01	0.01	0.06	3.0
Unpaved storage/parking	< 100	100	< 20	0.011	0.07	0.13	0.26	0.51	9.0
Impervious Areas:									
Sidewalks	< 50	< 50	< 20	< 0.005	0.11 0.02	0.05	0.09	0.47	3.7
Paved parking, storage, etc.	< 100	450	< 20	< 0.005	0.02	0.12	0.20	0.40	4.0
Road gutters	< 100	100	< 20	< 0.005	0.05		0.45	0.66	9.0
Residential/Commercial									
Pervious Areas:									
Grass/open areas	< 20	350	< 10	< 0.005	< 0.01	< 0.01	0.04	0.02	1.4
Impervious Areas:									
Sidewalks	75	600	< 20	< 0.005	< 0.01	0.02	0.15	0.16	1.4
Paved parking, storage, etc.	< 20	200	10	< 0.005	0.02	0.04	0.23	0.23	2.6
Paved roads	50	200	< 10	< 0.005	0.01	0.05	0.26	0.26	3.2
Road gutters	60	4,00	< 10	< 0.005	0.01	0.02	0.12	0.09	1.8
Roadside grass swales	60	1,00	< 10	< 0.005	< 0.01	0.01	0.05	0.08	1.6

[a]Fecal streptococca measured in number per 100 ml.

Source: Pitt and McLean, 1986.

- paved storage areas,
- unpaved parking and storage areas,
- driveways,
- streets,
- landscaped areas,
- undeveloped areas,
- freeway paved lanes and shoulders, and
- vehicle service areas.

Table 2.19 summarizes data describing urban area runoff pollutants from these source areas for different land uses and seasons. Snowmelt waters, especially from industrial source areas, are shown to be contaminated with many heavy metal pollutants, such as lead, zinc, and copper. Lead and zinc concentrations are generally the highest in sheet flows from paved parking areas and streets, with some high zinc concentrations also found in roof drainage samples. High bacterial populations have been found in sidewalk, road, and some bare ground sheet flow samples (collected from locations where dogs would most likely be "walked"). Bacterial levels are much lower during the cold season, but are still higher than most criteria allow.

Some of the sheet flow contributions observed at these locations were not sufficient to explain the concurrent concentrations of the same constituents observed in runoff at the outfall. The low chromium surface sheet flow concentrations at the Toronto industrial area occurring, at the same time as higher outfall chromium concentrations, indicated a high likelihood for direct industrial wastewater connections to the storm drainage system, for example. Similarly, most of the fecal coliform populations observed in sheet flows were also significantly lower than those observed at the outfall.

Treatment Methods

Stormwater runoff has been treated for reuse successfully in several U.S. cities. Many processes affect fate and removal mechanisms of pollutants in treatment facilities (Callahan et al., 1979). Sedimentation is the most common control mechanism for particulate-bound pollutants which typically are the constituents of concern in stormwater runoff. Exceptions include salt, zinc, 1,3-dichlorobenzene, fluoranthene, and pyrene, which may be mostly associated with the filterable sample portions of stormwater. Particulate removal can occur in many control processes, including catch basins, vegetation filtration, swirl concentrators, screens, drainage systems, and detention ponds.

Biological or chemical degradation of the toxicants may occur, but is quite slow for many of the pollutants in anaerobic environments. Degradation of soluble pollutants during treatment may occur, especially near the surface in aerated waters. Volatilization is also a mechanism that may affect many organic

TABLE 2.19 Birmingham Source Area Sheet Flow Quality

Pollutant (µg/l unless noted)	Residential Roofs		Commercial Roofs		Industrial Roofs		Resid./Inst. Streets		Industrial Streets		Resid. Pvd Parking		Com/Inst Pvd Parking	
	Total	Filtered	Total	Filtered	Total	Filtered	Total	Filtered	Total	Filtered	Total	Filtered	Total	Filtered
Microtox toxicity (I35, % light decreased)	41	--	29	--	20	--	9.5	--	34	--	37	--	30	--
pH (pH units)	6.5	--	6.2	--	8	--	7.2	--	7.9	--	7.1	--	6.8	--
Suspended solids (mg/l)	26.8	--	6.3	--	6	--	14.6	--	66.5	--	15.5	--	41	--
Turbidity (nephelometric turbidity units)	4.2	--	3.3	--	4	--	6.5	--	48.1	--	16.3	--	7.9	--
Particle size (median microns)	22.4	--	27	--	14	--	30	--	27	--	35	--	34.4	--
Heavy Metals														
Aluminum	1,810	362	152	82	319	167	181	165	4,520	1,260	2,500	610	568	94.4
Cadmium	6.4	0.14	0.43	0.08	1.2	0.47	0.46	0.31	56.7	0.25	35.3	0.2	2.6	0.5
Chromium	12.2	0.86	170		6.2	0.72	3	0.9	11	1.6	290		19.6	1.5
Copper	46	2.8	6.1	0.86	240	1.6	10	1.34	410	4.2	286	2.1	39.3	16.8
Lead	61		28.9		37.8	0.66	16.8	2.73	56.3	0.9	66.7	1	46.4	1.5
Nickel	15		24.4		4		2.23		19.9		35.3		33.1	2.8
Zinc	476	438	181	128	33.5	21.8	37.5	37.5	67.5	27.3	64	66.5	178	144
Base-Neutrals														
Bis (2-chloroethyl) ether	4.5		29.2	5.9	6.5				4.2	1			2.4	
1,3-Dichlorobenzene	14	3.8	29.8	7.9					1.6	1.1	21.4	6.1	8	2.5
Bis (chloroisopropil) ether	46		22.9								41		40.2	
Bis (2-chloroethoxy) methane														
Hexachloroethane			18.9	4.6	12.3	5.7							11.4	1.3
Napthalene	6.6		62.7										9.4	
Di-n-butyl phthalate														
Acenaphylene											20.8			
Fluorene													2	
Phenanthrene			7.7								1.9	1.9	3.1	
Anthracene			8.2								47.4	2.7	2.4	
Benzl butyl phthalate			36.2	1.9	6	3.8	153		0.5					
Fluoranthene			15.3								40.3	9.8	5.5	
Bis(2-ethyl hexyl)phthalate									0.6	0.6	28.3		2.5	
Pyrene			9.5						3.9		14.9			
Benzo(a) anthracene			5.8						4.2		66.4		3.8	
Chrysene			24.7						5.1		5.9		6.6	
Benzo(b) fluoranthene	4.4		89		7.3						39.4		6.5	
Benzo(k) fluoranthene	3.4		74.1		3.3						10.1			
Benzo(a) pyrene	9.3		100		13.3									
Benzo(g,h,i) perylene			100											
Pesticides														
alpha BHC					0.29									
delta BHC					0.39									
Aldrin					0.29									
DDT	9.4		0.2								0.23		0.36	
Chlordane	0.37		0.15		0.66				0.31		0.78	0.18	0.17	
Methoxychlor														
Endrin														

TABLE 2.19 (continued)

Pollutant (µg/l unless noted)	Inst. Unpvd Parking		Indus. Unpvd Parking		Com/Indus Pvd Storage		Indus Unpvd Storage		Loading Docks		Vehicle Service		Landscaped Areas	
	Total	Filtered	Total	Filtered	Total	Filtered	Total	Filtered	Total	Filtered	Total	Filtered	Total	Filtered
Microtox toxicity (135, % light decreased)	29	--	18	--	46	--	46	--	30	--	25	--	24	--
pH (pH units)	8.2	--	7.9	--	9.1	--	8.2	--	7.8	--	7.2	--	6.6	--
Suspended solids (mg/l)	391	--	170	--	15	--	152	--	40	--	23.8	--	37.6	--
Turbidity (nephelometric turbidity units)	392	--	42.4	--	10.2	--	81.7	--	16.5	--	10.2	--	65	--
Particle size (median microns)	38	--	42.7	--	47.3	--	23.8	--	31.7	--	37	--	29.2	--
Heavy Metals														
Aluminum	11,600	370	3,140	1,070	614	267	2,840	62.9	777	7.7	705	136	2,310	1,210
Cadmium		1.5	1		4.6	1.2	6.7	2.2	1.4	0.44	9.2	0.23	0.29	0.24
Chromium	1.5	2.3	6.6	1.9	33.8	0.3	100	9.4	17.1		74.3	0.9	94.4	1.8
Copper	390	0.9	13.3	3.6	16.7	0.9	468	306	21.6	6.9	135	6.8	94.4	3.3
Lead	66.6		28	1.7	21	1.7	165	17.8	65.1	1.1	63.4	1.3	28.5	0.7
Nickel	30	18	73.3		30.6		69	0.7	6.7	0.8	4.2	6.6	38.2	
Zinc	81.5		28.3	19	48	39.2	2,740	7.8	36.8	22.2	105	72.8	263	165
Base-Neutrals														
Bis (2-chloroethyl) ether	12.2										9.4	4.9	11.7	
1,3-Dichlorobenzene							3.6	3.3			28.8	10.6	13.5	2.6
Bis (chloroisopropl) ether											47.3		17.4	
Bis (2-chloroethoxy) methane													3.8	1.6
Hexachloroethane											11.9	11.1		
Napthalene											28.4	16.8	10.3	
Di-n-butyl phthalate											0.8	0.8		
Acenaphylene											0.6			
Fluorene											2.6		6	
Phenanthrene											9.3	2.6	4.5	
Anthracene											10.6	4.2	26	
Benzl butyl phthalate							1.3				16.9	1.7	8.2	0.7
Fluoranthene					10.7									
Bis(2-ethyl hexyl)phthalate							2				18.1	1.9	2.4	
Pyrene											14.3		11.2	
Benzo(a) anthracene											6.4			
Chrysene											39.7		6.3	
Benzo(b) fluoranthene											23.9		12.7	
Benzo(k) fluoranthene											36.2		11.2	
Benzo(a) pyrene														
Benzo(g,h,i) perylene														
Pesticides														
alpha BHC														
delta BHC														
Aldrin														
DDT							0.89		0.43		0.28			
Chlordane											0.18			
Methoxychlor														
Endrin														

Note: Average concentrations of observed compounds in micrograms per liter are given unless otherwise indicated.
Blanks indicate that all samples had nondetectable quantities; for heavy metals - 1 µg/l detection limits, except for 5 µg/l for aluminum and 0.1 µg/l for cadmium; for base neutrals - generally 1 µg/l detection limits; for pesticides - generally 0.3 µg/l detection limits

Source: Pitt et al., 1994.

toxicants. Increased turbulence and elevated oxygen levels promote these processes. Sorption of pollutants onto solids or metal precipitation increases the sedimentation potential of pollutants and also encourages more efficient bonding of the pollutants in soils, preventing their leaching to ground water. The following discussion summarizes observed pollutant removals obtained in stormwater control devices.

Sedimentation Treatment

Wet Detention Ponds and Artificial Wetlands. Detention ponds are probably the most common management practice for the control of stormwater runoff. If properly designed, constructed, and maintained, they can be very effective in controlling a wide range of pollutants. Artificial wetlands are being proposed as a method for stormwater control, especially in conjunction with wet detention ponds. However, performance data are extremely limited at this time.

There are many kinds of detention ponds, including dry ponds (which typically contain no water between storms), wet ponds (which contain standing water between storms), and combination ponds (which drain slowly after storms and may contain a small permanent pool). In a survey of cities in the United States and Canada, the American Public Works Association found more than 2,000 wet ponds (about half of which were publicly owned), more than 6,000 dry ponds, more than 3,000 parking lot multiuse detention areas, and more than 500 rooftop storage facilities (Smith, 1982).

Detention ponds have been required for some time in selected areas of the United States and are therefore more numerous in certain regions than in others. In Montgomery County, Maryland, for example, detention ponds were first required in 1971; more than 100 facilities were planned during that first year, and about 50 were actually constructed. By 1978, more than 500 detention facilities had been constructed in the county (Williams, 1982). In DuPage County, Illinois, near Chicago, more than 900 stormwater detention facilities (some natural) receive urban runoff (McComas and Sefton, 1985).

The Nationwide Urban Runoff Program (NURP) included full-scale monitoring of nine wet detention ponds (U.S. Environmental Protection Agency, 1983). About 150 storm events were comprehensively monitored at these ponds, and performances ranged from negative removals for the smallest up-sized pipe installation to more than 90 percent consistent removals of suspended solids at the largest wet ponds. The best ponds reported BOD and COD removals of about 70 percent, nutrient removals of about 60 to 70 percent, and heavy metal removals of about 60 to 95 percent.

Catchbasin and Sewer Cleaning. The mobility of catch basin sediments was investigated using particulate fluorescent tracers mixed with catch basin sediment (Pitt, 1979). The amount of sediment in catch basins (and on streets) and the sewer system at any time was large in comparison with individual storm runoff yields, but was not very mobile. Cleaning the material from catch basins reduces the potential of very large discharges during rare scouring rains and enables additional material to be captured.

Further research in Bellevue, Washington, investigated the accumulation rate of sediment in storm sewers and the effects of sewer cleaning on runoff discharges (Pitt, 1984). The main source of the sediment in the catch basins and the sewer system was found to be the street surfaces. A few unusual locations were dominated by erosion sediment originating from steep hillsides adjacent to the storm sewer inlets. The catch basin and sewer sediment consisted of the largest particles that were washed from the streets. Smaller particles that had washed from the streets during rains proceeded into the receiving waters, leaving behind the larger particles.

Catch basin sump particulates can be removed to eliminate this potential source of urban runoff pollutants. Cleaning catch basins twice a year was found to allow the catch basins to partially capture particulates for most rains. This cleaning schedule reduces the total solids and lead urban runoff yields by between 10 and 25 percent, and COD, total Kjeldahl nitrogen, total phosphorus, and zinc by between 5 and 10 percent (Pitt and Shawley, 1981; Pitt, 1984).

Fate of Pollutants in Sedimentation Facilities. The major fate mechanism in wet detention ponds, and in smaller sumps such as catch basins, is sedimentation. Unfortunately, sedimentation results in the accumulation of polluted sediments. These sediments can be anaerobic, with associated chemical and biochemical transformations. Resulting toxic chemical releases from heavily polluted sediments, plus the potential problems associated with the disposal of contaminated dredging spoils during maintenance, can present problems.

Other important fate mechanisms possible in wet detention ponds and wetlands, but probably not important in small sump devices, include volatilization and photolysis. Biodegradation, biotransformation, and bioaccumulation (into plants and animals) may also occur in ponds. Most wet detention ponds are completely flushed by moderate rains (probably every several weeks), depending on their design. Much of the runoff during moderate and large rains passes through the ponds over several hours during and immediately after rains. Sediments may reside in ponds for several to many years. Therefore, the time available for these other removal or transformation processes can vary greatly for different detention ponds. The residence time in small sedimentation devices is just a few minutes, and significant biological activity is not likely, except for some long-residing sediments that may become anaerobic in catch basin sumps.

Most sedimentation devices (especially ponds) are designed to provide effective sedimentation with sufficient storage for the long-term maintenance of accumulated sediment. The removal rates of many toxicants by other processes can possibly be increased by increasing water mixing, oxygen content, and biological activity in ponds.

Pollutant Removal During Infiltration

Grass filter strips can be quite effective in removing particulate pollutants from overland flows. The filtering effects of grasses, along with increased infiltration and recharge, reduce the particulate sediment load from urban landscaped areas. Grass filters can help reduce the particulate pollutant yields to the storm drainage system. Specific situations may include directing roof runoff to grass areas instead of pavement, planting grass between eroding slopes and the storm drainage system, and planting grass between paved or unpaved parking or storage areas and the drainage system.

Grass swale drainages are a type of infiltration device and can be used in place of concrete curb and gutters in most land uses, except possibly strip commercial and high density residential areas. Grass swales allow the recharge of significant amounts of surface flows while also providing some pollutant trapping in vegetation and surface soil. Because they are basically an infiltration device, swales should not be used in industrial areas because of the threat of ground water contamination.

Several large-scale urban runoff monitoring programs have included test sites that were drained by grass swales. For instance, one study of an area with poorly drained soils showed significantly lower surface flows (up to 95 percent lower) compared to a curb and gutter drained area (Bannerman et al., 1979). In another study, a special swale was constructed to treat runoff from a commercial parking lot. Soluble and particulate heavy metal (copper, lead, zinc, and cadmium) concentrations were reduced by about 50 percent. COD, nitrate nitrogen, and ammonia nitrogen concentrations were reduced by about 25 percent, while no significant concentration reductions were found for organic nitrogen, phosphorus, and bacteria (U.S. Environmental Protection Agency, 1983).

IRRIGATION RETURN FLOW

Characteristics

Irrigation return flow is the drainage water (surface and subsurface) collected from irrigated farmland. Because of the many types of crops irrigated, the wide variety of agricultural chemicals applied, the varying quality of supply water, and the different physical and chemical characteristics of soils, it is difficult to characterize the physical and chemical quality of irrigation return flow in any general way. Moreover, there is a paucity of data on the quality of irrigation return flow, except for data on dissolved solids and nitrate concentrations. Recent studies have addressed the content of selected pesticide residues and trace

Stormwater Infiltration at Fresno, California

The Nationwide Urban Runoff program (NURP) (U.S. Environmental Protection Agency, 1983) was conducted in the early 1980's to gather data on stormwater pollutant quality and quantity throughout the nation and to obtain information concerning the effectiveness of different stormwater management practices (Harrison, 1984). One site selected for study was Fresno, California. Fresno had 74 recharge basins, with a total infiltration area of over 1,000 acres. The basins provided local stormwater drainage and helped recharge the local aquifer (which is a sole-source aquifer). Five basins were studied; three were mostly grass-lined, while two were mostly bare dirt-lined. Two of the basins had been in operation for over 20 years. Stormwater from four land uses (medium-density residential, high-density residential, commercial shopping center, and moderate industrial) was sampled. Pollutant analyses were conducted on rainfall, dry atmospheric deposition, street dirt, stormwater, recharge basin soils, vadose water, and ground water.

Industrial stormwater runoff was the most severely polluted and fluctuated greatly during the storms. Runoff quality from other land uses, as expected, had the worst quality during the beginning of storms and early in the rainy season. The study found that soils in the recharge basins provided a high degree of removal of most of the stormwater pollutants. A correlation was found between the degree of metal removal and the proportion of silt plus clay and organic matter in the surface soils. The only organic compound found in the ground water beneath the basins was phenol, while basin soils were contaminated with lindane, chlordane, DDE, PCBs, and phenols. Arsenic, chromium, copper, lead, mercury, nickel, and zinc also were found in the ground water beneath the basins and in the basin soils. Lead and chlordane were the most commonly detected soil contaminants. Some of these pollutants also were found in the regional ground water. Other pollutants of potential concern in studies of ground water contamination by stormwater, but not monitored during the Fresno project, include bacterial and viral pathogens.

Although there was evidence of downward movement of some pollutants in the soils, the study concluded that there were no apparent adverse effects on ground water quality from infiltrating stormwater, especially with cleaner water being used for infiltration during dry months. The stormwater runoff had better mineral quality (as indicated by specific conductance) and lower nutrients than the regional ground water.

This study did not investigate recharge basins receiving mostly industrial waters because the researchers felt that more data would be needed before industrial water recharge effects could be known. There was some concern that lead might create problems in basins used for recreation, but it was felt that the exposure would be reasonably low. As a precaution, periodic replacement of turf or soil (with proper disposal) could be conducted in basins having recreational uses to reduce potential exposure and the possibility of leaching of contaminants from the soil in future years.

elements in irrigation return flow, but the data are incomplete and comprehensive summary assessments are not yet available. From the information available, the principal water quality variables of irrigation return flow with respect to its possible use as a source for ground water recharge are dissolved salts, fertilizer and/or pesticide residues, trace elements leached from soils, and suspended-solids load.

Drainwater from irrigated fields historically has been known to affect the quality of water in receiving streams, reservoirs, and wetlands by increasing concentrations of dissolved solids and major constituents (Engberg et al., 1991). In the last decade or so, attention has been directed toward trace constituents in irrigation drainage because of the linkage made between elevated concentrations of selenium in drainwater and the damage to the waterfowl and shorebird population at Kesterson Reservoir in California. In 1985 the Department of the Interior (DOI) initiated the National Irrigation Water Quality Program, a five-phase research program to identify water quality problems caused by irrigation drainwater, through reconnaissance and detailed studies, and to plan and implement remediation actions. The reconnaissance study phase involved sampling of water, bottom sediment, and biota at 26 areas in western states, before, during, and after the irrigation season (Engberg et al., 1991). Samples of each medium were analyzed for major constituents; trace elements including arsenic, barium, boron, cadmium, chromium, copper, lead, mercury, molybdenum, nickel, selenium, silver, uranium, vanadium, and zinc; and pesticide residues in some cases. These studies should provide a consistent database useful for characterizing the quality of irrigation drainwater, but currently only limited summary results are available in the literature.

Overall, research at the seven reconnaissance sites showed that selenium was the constituent most frequently detected at elevated concentrations in wetland ecological systems (Deason, 1989). Also, the concentrations of constituents were found to vary widely on a spatial basis and, therefore, irrigation induced contamination problems are likely to be very site specific. In addition to the selenium, boron, arsenic, uranium, and mercury found at elevated concentrations, one pesticide residue, DDE, was frequently found at elevated concentrations (Feltz et al., 1990). Closed drainage basins, especially terminal ponds, wetlands, and playas, had the highest concentrations of dissolved solids and specific constituents of concern.

The great variation in irrigation return flow quality can be seen by inspecting results reported from several widely scattered sites in the western United States. From a study of return flow from the Milk River Irrigation Project in northeastern Montana, Lambing et al. (1988) concluded that, "Results of the current study indicate that irrigation drainage sampled in 1986 had relatively small concentrations of most constituents. The only significant differences between the supply water (Milk River) and irrigation drainage were zinc (56 µg/l) and uranium (13 µg/l) concentrations that were several times larger in one of the irrigation drains."

For another DOI study site, the Angostura Reclamation Unit in southwestern South Dakota, Greene et al. (1990) reported that "irrigation return flow had relatively small concentrations of trace elements. Overall, there appeared to be minor differences between concentrations of trace elements in water of the Cheyenne River upstream of irrigated land and in water downstream from all irrigation return flow."

For yet another DOI site, the Fallon agricultural area in west central Nevada, Hoffman et al. (1990) reported that irrigation drainwater had a specific conductance that ranged from 566 to 41,000 microsiemens per centimeter (µS/cm) at 25°C with a median of 1,990 µS/cm. In contrast, the source water had a conductance that ranged from about 200 to 400 µS/cm with a median of about 250 µS/cm. These data indicate an eight to tenfold increase in dissolved solids in irrigation drainwater compared to the source water. With respect to trace elements, they report that dissolved aluminum, barium, cadmium, chromium, copper, lead, lithium, molybdenum, nickel, silver, and vanadium were present in concentrations either below Nevada criteria for the protection of aquatic life or propagation of wildlife, or below the analytical reporting level. On the other hand, arsenic (range 1 to 190, median 44 µg/l), boron (range 190 to 28,000, median 2,200 µg/l), and uranium (up to 300 µg/l) were notably high in irrigation drainwater. Although some enrichment of nitrogen and phosphorus occurred during the irrigation process, drainwater concentrations of these constituents were typically less than 1 mg/l.

Other studies conducted in California and Colorado further illustrate the variability of irrigation return flow. Tanji (1981) reported on irrigation return flow quality for the Glenn-Colusa Irrigation District and the Panoche Drainage District in the northern and central part of the Central Valley of California, respectively. He showed an increase in total dissolved solids (TDS) of 2 to 10 times, suspended solids of 1.5 to 5 times, nitrate-nitrogen of 1.3 to 20 times, and boron of almost 40 times over the concentrations in the supply water. The constituent concentrations in the irrigation return flow were as high 2,050 mg/l for TDS, 348 mg/l for suspended solids; nitrate-N: 12.6 mg/l for nitrate-nitrogen; and 4.3 mg/l for boron.

Keys (1981) reported the irrigation return flow quality contrasted with the supply water quality for the Grand Valley area in west central Colorado. TDS concentrations remained about the same in the surface irrigation return flow as in the supply water (about 400 mg/l), but increased by about 10 times (to 4,100 mg/l) in the subsurface irrigation return flow water. Nitrate-nitrogen increased from 0.3 mg/l in the source water to 4 mg/l in the combined surface and subsurface irrigation return flow. Phosphates did not change in the return flow, but remained the same as in the supply water (0.05 mg/l).

Both the California and the Colorado cases indicate a significant elevation in constituent concentrations or physical property in the return flow compared to the supply water, especially for the subsurface drainage. However, the extent of enrichment through the irrigation process and the concentrations in the drainwater differ widely.

Analysis of San Luis drainwater in the San Joaquin Valley of California indicated the drainwater there to be high in dissolved solids (about 10,000 mg/l), sulfate (about 5,000 mg/l), sodium (about 2,200 mg/l), and chloride (about 1,500 mg/l) (Lee, 1990). Of the trace elements, boron (14.4 mg/l) and selenium (0.3 mg/l) are notably high. Nitrate/nitrite as N was also high, averaging 48 mg/l. The average and maximum concentrations of various constituents in the drainwater are given in Table 2.20.

Agricultural drainage water from the San Joaquin Valley contained an average of 20 mg/l of nitrate-nitrogen but some areas had concentrations of 100 to 200 mg/l of (Bouwer, 1987). Total concentrations of other forms of nitrogen rarely exceeded 1 mg/l. The average concentration of phosphorus was 0.09 mg/ and salt (TDS) was 3,625 mg/l. The average concentrations of nitrate-nitrogen were 19 mg/l, and TDS 3,600 mg/l for tile drainage systems in the San Joaquin Valley for the period from 1962 to 1969 (Schmidt and Sherman, 1987).

Research on pesticide concentrations in surface irrigation runoff water and in tile drain effluents following application of pesticides to large fields of cotton, sugarbeets, alfalfa, lettuce, onions, and cantaloupes in the Imperial Valley, California, found that concentrations of pesticides in surface runoff water were dependent on the characteristics of the pesticides, their methods and rates of applications, the time elapsed between application and the first irrigation, the number of irrigation cycles since the pesticide application, irrigation efficiency, and other soil management practices (Spencer et al., 1985). Seasonal totals of insecticide in surface runoff were less than 1 percent of the amount applied, whereas herbicides in surface runoff were usually 1 to 2 percent of the amounts applied. None of the pesticides were identified in tile drain effluents at concentrations above minimum detectable levels of 1 to 2 ng/l. The highest mean concentrations calculated for the various pesticides in irrigation runoff water are listed in Table 2.21.

Several studies investigated pesticide residues in surface and tile drainage water (Pierce and Wong, 1988). These studies indicated (1) atrazine residue mean concentrations of about 14 to 56 µg/l and maximum of 1,100 µg/l from tailwater pits in Kansas cornfields; (2) alachlor, cyanazine, propazine, and terbutryne present in water and sediment from the same tailwater pits; (3) terbacil and 2,4-D at concentrations of 10 and 110 µg/l, respectively, in drainage water from citrus groves; (4) residues of the triazine herbicides, atrazine, cyanazine, cipiazine, and metabuzine in tile drainwater from Quebec cornfields; and (5) atrazine residues in tile drainwater from Ontario cornfields.

Leaching of nitrate and, to a lesser extent, pesticides to ground water beneath irrigated fields has been observed in many states (Law, 1987; Sabol et al.,

TABLE 2.20 Drainage Water Analysis for the San Luis Drain, Mendota, California

Constituent	Units	Average	Maximum
Sodium	mg/l	2,230	2,820
Potassium	mg/l	6	12
Calcium	mg/l	554	714
Magnesium	mg/l	270	326
Alkalinity (as calcium carbonate)	mg/l	196	213
Sulfate	mg/l	4,730	6,500
Chloride	mg/l	1,480	2,000
Nitrate/nitrite (as N)	mg/l	48	60
Silica	mg/l	37	48
Total dissolved solids	mg/l	9,820	11,600
Suspended solids	mg/l	11	20
Total organic carbon	mg/l	10.2	16
Chemical oxygen demand	mg/l	32	80
Biochemical oxygen demand	mg/l	3.2	5.8
Temperature[a]	°C	19	29
pH		8.2	8.7
Boron	µg/l	14,400	18,000
Selenium	µg/l	325	420
Strontium	µg/l	6,400	7,200
Iron	µg/l	110	210
Aluminum	µg/l	< 1	< 1
Arsenic	µg/l		
Cadmium	µg/l	< 1	20
Chromium (total)	µg/l	19	30
Copper	µg/l	4	5
Lead	µg/l	3	6
Manganese	µg/l	25	50
Mercury	µg/l	< 0.1	< 0.2
Molybdenum	µg/l	88	120
Nickel	µg/l	14	26
Silver	µg/l	<1	<1
Zinc	µg/l	33	240

[a]Temperature varied from 23 to 25°C (summer) to 12 to 1 5°C (winter).

Source: U.S. Bureau of Reclamation, 1985, after Lee, 1990.

TABLE 2.21 Highest Mean Concentration Calculated for Pesticides in Irrigation Runoff Water for Eight Fields in Imperial Valley, California

	Highest Mean Concentration µg/l
Herbicides	
Cycloate	2.5
DCPA	153
Dinitramine	15
EPTC	1,250
Prometryn	180
Trifluralin	10.7
Insecticides	
Organophosphates	
Azinphosmethyl	ND
Chloropyrifos	22.4
Diazinon	8.6
Malathion	13.3
Methidathion	64
Mevinphos	1.26
Ethyl parathion	50
Methyl parathion	18
Sulprofos	1.8
Carbamates	
Methomyl	119
Organochlorines	
Endosulfan	71
Ethylan	2
Pyrethroids	
Fenvalerate	5.1
Permethrin	3.4

Note: ND = not detected.

Source: Spencer et al., 1985.

1987; Sonnen, et al., 1987; Mossbarger and Yost, 1989; Ritter et al., 1989, 1991). Nitrate contamination of ground water is especially prevalent in irrigated areas (Power and Schepers, 1989). Irrigation return flow collected by tile drains, therefore, should be expected to contain elevated nitrate concentrations and perhaps some level of pesticide contamination, depending on the type of pesticide applied.

Irrigation return flow can also contain wide ranging concentrations of suspended sediment, as shown by Boucher (1984) for a study of irrigated land in south central Washington. Discharge-weighted mean concentrations of suspended sediment for four drain outflows ranged from 7 (±3) to 1,390 (±100) mg/l for the 1980 irrigation season and from 9 (±1) to 2,800 (±200) mg/l for the 1981 irrigation season. Suspended solids concentrations in irrigation return flow for other areas are listed in analyses presented earlier and also show wide variations.

Irrigation return flow constitutes a large supply of water nationally, but especially in the semiarid West. Its physical and chemical quality, however, varies widely. Typically, irrigation return flow contains high concentrations of dissolved solids. It also may contain objectionable concentrations of nitrate, pesticide residues, and trace elements. The suspended solids load of irrigation return flow may also be high. Collectively, these quality characteristics generally indicate the need for extensive treatment of irrigation return flowbefore it might be used for recharge without causing quality problems in the receiving ground water.

Treatment Methods

Water quality parameters of greatest concern in irrigation return flow include suspended solids, TDS, nitrogen and phosphorus compounds, pesticide residues, and various trace metals. Treatment technologies to remove each of these pollutants are available.

Suspended solids can be removed by settling ponds, chemical clarification, filtration, and membrane processes. These methods are sometimes used in series to produce a highly clarified end product free of turbidity and very low in suspended solids concentration. For surface infiltration systems, only settling ponds may be required; however, if the water is to be recharged by wells directly into the aquifer, then chemical clarification, filtration, and membrane processes are needed to produce a suitable effluent. These processes are commonly used to treat surface waters intended for water supply and to treat municipal wastewater prior to disposal or ground water recharge.

Irrigation return flow most often has a TDS concentration that renders it undesirable for recharge without treatment to reduce the mineralization. Reverse osmosis can be used to reduce TDS levels regardless of their initial concentration. For instance, roughly 90 to 95 percent of the inorganics and 95 percent of dissolved organics can be removed by a well-maintained reverse osmosis facility (Treweek, 1985). In addition, particulate matter, bacteria, and viruses also are removed, although the latter are generally not of great concern in irrigation return flow. Although reverse osmosis is widely used in the treatment of brackish and saline waters for water supply, it probably would be economically prohibitive for irrigation return flows.

Nitrate, the dominant nitrogen species in irrigation return flow, can be reduced to acceptable levels by reverse osmosis, if overall water composition requires such treatment. If it is unnecessary to treat the water for TDS reduction, then nitrate levels can be reduced instead by biological denitrification, a process that has been applied in the past to treat municipal wastewater effluent. Phosphorus compounds are readily removed by chemical clarification, as are most of the trace metals commonly found in irrigation return flow.

Granular activated carbon (GAC) adsorption can be used if needed to remove soluble pesticide residues remaining after other treatment processes are completed. GAC is a commonly used treatment method for removing dissolved organics in drinking water supplies and has also been used to treat municipal wastewater (Treweek, 1985).

SUMMARY

Wastewaters considered suitable source waters for ground water recharge include municipal wastewater effluent, stormwater runoff, and irrigation return flow. Table 2.22 is a summary of advantages and disadvantages pertaining to the use of wastewaters for recharge, Table 2.23 is a summary of the qualities of the three primary types of wastewater considered in this report. Of the three, treated municipal wastewater effluent is by far the most consistent, spatially and temporally, in both quantity and quality. An exception to this generalization is where raw municipal wastewater and stormwater are commingled in a combined sewerage system.

When compared to other potential impaired water sources, the quality of treated municipal wastewater has been characterized extensively for various levels of treatment because of regulations pertaining to the disposal of municipal wastewater effluent and because municipal wastewater has a history of use as a recharge water source. The body of information on quality of stormwater runoff and irrigation return flow is far less developed, especially when their greater variability is taken into account. Therefore, characterization of stormwater runoff and irrigation return flow quality must be drawn from a much less systematic and comprehensive database than is available for municipal wastewater.

Constituents of concern in municipal wastewater include organic compounds, nitrogen species, pathogenic organisms, and suspended solids. Treatment processes are readily available and have been used successfully to treat municipal wastewater effluent to levels acceptable for various recharge applications. However, even when treated to a very high degree, disinfection of the effluent with chlorine results in the formation of disinfection by-products with the residual organic compounds. These DBPs are of concern if the recovered ground water is to be used for potable purposes.

Urban stormwater runoff quality is affected by several factors, including rainfall quantity and intensity, the natural and anthropogenic characteristics of the drainage basin, time since the last runoff event, and, in northern areas, the time of year. Constituents of concern in urban stormwater runoff include trace metals, organic compounds, pathogenic organisms, suspended solids, and in northern climates in the winter, dissolved solids and chloride enrichment by road deicing practices. Stormwater runoff typically is not treated. However, field experience suggests that many suitable treatment methods exist that may adequately treat most stormwaters before surface infiltration. Overall, stormwater from residential areas is generally best in quality, but its quantity may be extremely erratic and unpredictable due to natural rainfall variations. Recharge with stormwater often requires surface storage and flow regulation because recharge systems do not have the capacity to allow immediate infiltration all the runoff produced by a given precipitation event. Residential area stormwater runoff is best allowed to infiltrate through source area recharge devices, such as french drains, grass filter strips, and grass drainage swales.

Although not examined in depth in this report, industrial stormwater runoff is very irregular in quality, especially for toxicants. Because of this irregular quality and the great potential for severe contamination, industrial area stormwater runoff is not a good candidate for ground water recharge use. Urban snowmelt may also be a poor choice for recharge because of its high salt content. Dry weather flow in stormwater drainage systems may be associated with highly contaminated inappropriate discharges (such as raw municipal wastewater, industrial process water, and illegal dumping of hazardous materials) and should also be avoided. Therefore, to take advantage of urban stormwater runoff as a source of recharge water, care must be taken to isolate the acceptable residential area runoff from the more contaminated flows or to provide additional source area treatment for runoff from the critical areas.

Irrigation return flow exhibits the widest variation in quality of the three potential source waters. It varies from having basically the same quality as high-quality surface water to having a salinity of as much as 10,000 mg/l. The quality characteristics of irrigation return flow are not well studied, except for salinity and concentrations of nitrate. In humid areas, the salt content of irrigation return flow is not a problem, but in semiarid areas it can be enriched to 8 to 10 times that of the water applied. Nitrate concentrations can be as high as 100 to 200 mg/l. Suspended solids and trace element concentrations including selenium, uranium, boron, and arsenic are also of concern. Pesticide residues may also pose problems in irrigation return flow, but in general most of the pesticide residues are associated with particulates and are readily removed with suspended solids. Treatment of irrigation return flow is not generally done, but treatment processes are available to remove the constituents of concern to acceptable levels. The cost-effectiveness of doing so for saline waters is questionable.

In the past, surface and subsurface return flows from irrigated agricultural areas were simply "disposed" of in streams, lakes, and the ocean without any environmental concerns. This attitude is changing and there is a trend toward increased management to minimize degradation of the environment, such as storage in evaporation ponds and ultimate disposal of salts as solid waste, treat-

TABLE 2.22 Advantages and Disadvantages for Using Various Wastewaters in Ground Water Recharge

	Advantages	Disadvantages
Municipal Wastewater		
Primary-treated municipal wastewater	• High TOC for possible improved denitrification • Relatively constant flow • Located near major point of use	• Poor water quality; higher toxicants, nutrients, BOD, and suspended solids than other municipal wastewaters • High disinfection by-product formation potential
Secondary-treated municipal wastewater	• Most common • Relatively constant flows • High volume • Located near major point of use	• Moderate to poor water quality
Advanced treated municipal wastewater	• Best quality municipal wastewater • Low TOC for reduced disinfection by-product formation potential • Relatively constant flows • Located near major point of use	• High cost
Agricultural Irrigation Return Flows		
Irrigation return flow		• High pesticides and herbicides • High nutrients and salts • Irregular flows

Urban Stormwater

Residential area stormwater
- Likely best quality wastewater
- Most common stormwater
- Located near major point of use

Industrial area stormwater
- Highly irregular toxicant quality (likely contamination from industrial processes and contact with grossly polluted soils)

Urban snowmelt water
- High salt content in areas using common de-icing procedures

Dry weather stormwater sewerage flows
- High pesticides and herbicides
- Likely contamination from inappropriate discharges

Combined Sewage

Combined sewage
- contains raw sewage with pathogen contamination
- higher likelihood of toxicants from stormwater from older industrial and commercial areas

Irregular flows (highly intermittent)

TABLE 2.23 Comparison of Quality Parameters for Irrigation Return Flow, Urban Stormwater Runoff, and Treated Municipal Wastewater

	Irrigation Return Flow	Urban Stormwater	Secondary Treated Municipal Wastewater
General	• TDS > 500-10,000 mg/l. (There is extreme variability; low TDS is largely Ca^{++}, Mg^{++}, HCO_3^-, and high TDS largely Na^+, Cl^-, SO_4^{--}, depending on source quality and location.) • Total suspended solids 7-3,000 mg/l • Nitrate, up to 10-20 times drinking water standards	• Turbidity: 10-100 Nephalometric turbidity units (>1,000 NTU in construction area runoff) • Total suspended solids: 25-1,000 mg/l (>10,000 mg/l in construction runoff) • Total dissolved solids: 50-100 mg/l (Cl can be 10,000 mg/l from applied salt in northern areas) • Chemical oxygen demand 50-100 mg/l • BOD_5: 10-25 mg/l (ultimate biochemical oxygen demand in several hundred mg/l) • Nitrate: 0.5-10 mg/l	• Total suspended solids: 10-25 mg/l • Total dissolved solids: 200-5,000 mg/l • Total organic carbon: 15-25 mg/l • Chemical oxygen demand: 40-70 mg/l • BOD_5: 15-30 mg/l • Nitrate: 0.4-30 mg/l
Biological	• Inadequate data	• Human pathogens: *Pseudomonas aeruginosa*: 10^3-10^5/100 ml *Shigella*, protozoa, and viruses: likely present (limited data) • Fecal coliforms: 10^3-10^5/100ml	• Human pathogens: Almost complete removal with disinfection • Fecal coliforms: < 10/100 ml
Organics	• No extensive data currently available • Pesticides, dependent on local practice; some notable concentrations observed: terbacil: 10 μg/l 2,4-D: 110 μg/l atrazine: < 10-1,000 μg/l chlorothanlonil: 0.04-3.7 μg/l DCPA: 150 μg/l (mean) EPTC: 1,250 μg/l (mean) Prometryn: 180 μg/l (mean) Methomyl: 120 μg/l (mean)	• Pesticides, mostly in residential base flows: DDT: up to 1 μg/l lindane: up to 1 μg/l endrin: up to 1 μg/l chlordane: up to 10 μg/l methoxychlor: up to 10 μg/l • Volatile organic compounds (benzene and toluene present in industrial runoff) • Polycyclic aromatic hydrocarbons (mostly in particulate forms): benzo(a)pyrene: 10-100 μg/l fluoranthene, phenanthrene, chrysene, pyrene, and anthracene: 1 to 25 μg/l • bis(2-ethylhexyl)phthalate: up to 60 μg/l • Others: 1,2-dichloroethane, ethylbenzene pentachlorophenol, and tetrachloroethylene: up to 10 μg/l (mostly in industrial runoff) PCBs: up to 10 μg/l (industrial runoff)	• Volatile organic compounds: methylene chloride: < 1-5 μg/l chloroform: 3.5-7.5 μg/l • Polycyclic aromatic hydrocarbons: benzo(a)pyrene: 0.1-0.4 μg/l fluoranthene: 0.01-0.35 μg/l pyrene: 0.2-3 μg/l • Phthalates: up to 200 μg/l • Others ethylbenzene: 11 μg/l pentachlorophenol: 4 to 300 μg/l 1,1,2,2-tetrachloroethylene: 1-60 μg/l
Metals and Other Inorganics	• Mostly associated with soluble phase • Selenium: up to 300 μg/l • Uranium: up to 300 μg/l • Boron: 190-28,000 μg/l • Arsenic: 1-190 μg/l	• Mostly associated with particulate phase • Antimony: 2-100 μg/l • Chromium: 1-200 μg/l 25% filterable (mostly in industrial runoff) • Arsenic: 1-50 μg/l (average 1 μg/l) • Beryllium: 1-50 μg/l • Cadmium: 1-15 μg/l (25% filterable) • Selenium: 1-100 μg/l • Zinc: 200 μg/l average, can be 1,000 μg/l in runoff from galvanized metal (mostly filterable) • Mercury: 0.1-200 μg/l • Nickel: 1-200 μg/l (50% filterable) • Cyanide: 1-300 μg/l • Lead: 10-500 μg/l, has decreased by about 10 $\times$ over past 20 years (only 10% filterable)	• Arsenic: 5-23 μg/l • Boron: 300-2,500 μg/l • Cadmium: < 5-200 μg/l • Chromium: < 1-100 μg/l • Copper: 6-50 μg/l • Lead: 3-350 μg/l • Mercury: < 2-10 μg/l • Nickel: 3-600 μg/l • Zinc: 4-350 μg/l

ment, or deep-well injection into closed geologic formations. To minimize the cost of these techniques, irrigation efficiencies must be increased, and return flows must be reused as much as possible for irrigation (e.g., raising salt tolerant crops) to minimize the volume of irrigation return flow ultimately produced.

The availability of wastewater for recharge can vary widely throughout the country. Total urban stormwater runoff may provide about 100 acre-feet per square mile per year in areas having about 25 cm (10 inches) of rain per year, but can increase to about 1,000 acre-feet per square mile per year in more humid areas of the United States (having about 125 cm (50 inches) of rain per year). Obviously, some of this water would be diverted from recharge facilities because of poor quality (such as that from industrial areas), reducing the amount available. Municipal wastewater flow would also vary, depending on the population served in a community. Large cities may provide as much as 1,500 acre-feet per square mile per year of wastewater for recharge, but most small towns will generate only about one-tenth as much because of lower population densities. The amount of irrigation return flow available for recharge would vary greatly, depending on irrigation practice, return flow collection efficiency, crop requirements, and rainfall amounts. The amount of irrigation return flow to surface or ground water supplies in 1985 in the United States was estimated to be about 45 million acre-feet (Solley et al., 1988). This amount is 29 percent of the 154 million acre-feet withdrawn for irrigation in 1985. California and Idaho were by far the largest users of irrigation water, together accounting for 37 percent of the national total.

Finally, location of the wastewater source is also important. While urban stormwater runoff and municipal wastewater are usually located near the area of use, most irrigation return flows would be located further from populated areas.

REFERENCES

American Society of Civil Engineers. 1970. Engineering evaluation of virus hazard in water. J. Sanit. Eng. Div., 96(SA1):111.

Amy, G. L., P. A. Chadki, and P. H. King 1984. Chlorine Utilization during formation of THM in presence of ammonia and bromide. Environ. Sci. Technol. 18:781-786.

Bannerman, R., J. G. Konrad, and D. Becker. 1979. The IJC Menomonee River Watershed Study. EPA-905/4-79-029. U.S. Environmental Protection Agency. Chicago, Ill.

Battigelli, D. A., M. D. Sobsey, and D. C. Lobe. 1993. The inactivation of hepatitis A virus and other model viruses by UV irradiation. Water Sci. Technol. 27(3-4):339-342.

Bookman-Edmonston Engineering, Inc. 1992. Annual Report on Results of Water Quality Monitoring: Water Year 1990-91. Report prepared for the Water Replenishment District of Southern California by Bookman-Edmonston Engineering, Inc., Glendale, California.

Boucher, P. R. 1984. Sediment transport by irrigation return flows in four small drains within the DID-18 drainage of the Sulphur Creek basin, Yakima County, Washington, April 1979 to October 1981: U.S. Geol. Surv. Water Resour. Invest. Rep. 83-4167, 149 pp.

Bouwer, H. 1987. Effect of irrigated agriculture on groundwater. J. Irrig. Drain. Eng. 113(1):4-15.

Bull, R. J., and F. C. Kopfler. 1991. Health Effects of Disinfectants and Disinfection By-products. Denver: Am. Water Works Assoc. Res. Found.

Callahan, M. A., M. W. Slimak, N. W. Gabel, I. P. May, C. F. Fowler, J. R. Freed, P. Jennings, R. L. Durfee, F. C. Whitmore, B. Maestri, W. R. Mabey, B. R. Holt, and C. Gould. 1979. Water Related Environmental Fates of 129 Priority Pollutants, EPA-4-79-029a and b. Monitoring and Data Support Division, U.S. Environmental Protection Agency, Washington, D.C.

Carlson, R. R., K. D. Lindstedt, E. R. Bennett, and R. B. Hartman. 1982. Rapid Infiltration Treatment of Primary and Secondary Effluent. J. Water Pollut. Contr. Fed. 54(3):270-280.

Casson, L. W., C. A. Sorber, J. L. Sykora, P. D. Gavaghan, M. A. Shapiro, and W. Jakubowski. 1990. Giardia in Wastewater: Effect of Treatment. Res. J. Water Pollut. Contr. Fed. 62(5):670-675.

Cavanaugh, J., H. Weinberg, A. Gold, R. Sangaiah, D. Marburg, W. H. Glaze, T. Collete, S. Richardson, and A. Thruston. 1992. Ozone by-products: Identification of bromohydrins from the ozonation of natural waters with enhanced bromide levels. Environ. Sci. Technol. 26(8):151-157.

CH2M Hill. 1992. UV Disinfection Pilot Study: Rapid Infiltration/Extraction (RIX) Demonstration Project. Prepared for the Santa Ana Watershed Project Authority, City of San Bernardino, and City of Colton. Santa Ana, Calif.

CH2M Hill. 1993. Advanced Wastewater Reclamation Program Final Report. Prepared for the Reedy Creek Improvement District, Lake Buena Vista, Florida, by CH2M Hill, Gainesville, Florida.

Chen, C., R. Nur, J. F. Stahl, R. W. Horvath, and J. F. Kuo. 1993. UV inactivation of bacteria and viruses in tertiary effluents. In Proceeding Specialty Series: Training design and operation of effluent disinfection systems, May 23-25, Hanover, NJ. County Sanitation Districts of Los Angeles County, Whittier, Calif.

Colclough, C. A., J. D. Johnson, R. F. Christman, and D. S. Millington. 1983. Environmental Impact and Health Effects. Pp. 219-229 in Water Chlorination: Chemistry, Environmental Impact and Health Effects, Vol. 4, R. L. Jolly et al., eds. Ann Arbor, MI: Ann Arbor Science.

Crook, J. 1992. Water reclamation. Pp. 559-589 in Encyclopedia Physical Science and Technology, Vol 17. San Diego, Calif.: Academic.

Deason, J. P. 1989. Irrigation-induced contamination: How real a problem. J. Irrig. Drain. Eng. 115(1):9-20.

Dizer, H., W. Bartocha, H. Bartel, K. Seidel, J. M. Lopez-Pila, and A. Grohmann. 1993. Use of ultraviolet radiation for inactivation of bacteria and coliphages in pretreated wastewater. Water Res. 27(3):397-403.

Engberg, R. A., M. A. Sylvester, and H. R. Feltz. 1991. Effects of drainage on water, sediment, and biota. Pp. 801-807 in Proceedings of the 1991 National Conference Sponsored by the Irrigation and Drainage Division of the American Society of Civil Engineers, W. F. Ritter, ed. July 22-26, 1991, Honolulu, Hawaii.

Environment Canada. 1980. Rideau River Water Quality and Stormwater Monitor Study, 1979. MS Rep. No. OR-29 (Feb). Hull, Quebec.

Feachem, R. G., D. J. Bradley, H. Garelick, and D. D. Mara. 1983. Sanitation and Disease—Health Aspects of Excreta and Wastewater Management. Chichester, England: John Wiley, for the World Bank.

Feltz, H. R., R. A. Engberg, and M. A. Sylvester. 1990. Investigations of water quality, bottom sediment, and biota associated with irrigation drainage in the western United States. Pp. 119-130 in The Hydrological Basis for Water Resources Management: Proceedings of the Beijing Symposium, Oct. 1990. IAHS Publ. No.197.

Field, R., and M. O'Shea. 1993. The detection and disinfection of pathogens in storm-generated flows. Water Sci. Technol. 28(3-5):311-315.

Field, R., V. P. Olivieri, E. M. Davis, J. E. Smith, and E. C. Till, Jr. 1976. Proceedings of Workshop on Microorganisms in Urban Stormwater. EPA-600/2-76-244. Office of Res. and Develop., U.S. Environmental Protection Agency. Cincinnati, Ohio.

Galvin, D. V., and R. K. Moore. 1982. Toxicants in Urban Runoff. Toxicant Control Planning Section, Municipality of Metropolitan Seattle. Contract P-16101. U.S. Environ- mental Protection Agency. Lacy, Wash.

Geldreich, E. E., and B. A. Kenner. 1969. Concepts of fecal streptococci in stream pollution. J. Water Pollut. Contr. Fed. 41(8):R336-R352.

Glaze, W. H. 1986. Reaction products of ozone: a review. Environ. Health Persp. 69:151-157.

Glaze, W. H., J. B. Andelman, R. J. Bull, R. B. Conolly, C. D. Hertz, R. C. Hood, and R. A. Pegram. 1993. Determining health risks associated with disinfectants and disinfection by-products: research needs. J. Am. Water Works Assoc. 83(Feb.):3.

Greene, E. A., C. L. Sewards, and E. W. Hansmann. 1990. Reconnaissance investigation of water quality, bottom sediment, and biota associated with irrigation drainage in the Angostura Reclamation Unit, southwestern South Dakota, 1988-89: U.S. Geol. Surv. Water Resour. Invest. Rep. 90-4152, 75 pp.

Gupta, M. K., R. W. Agnew, D. Gruber, and W. Kreutzberger. 1981. Constituents of Highway Runoff. Vol IV: Characteristics of highway runoff from operating highways. Fed. Highway Admin. Rep. No. FHWA/RD-81/045.

Hoffman, R. J., R. J. Hallock, T. G. Rowe, M. S. Lico, H. L. Burge, and S. P. Thompson.1990. Reconnaissance investigation of water quality, bottom sediment, and biota associated with irrigation drainage in and near Stillwater Wildlife Management Area, Churchill County, Nevada, 1986-87. U.S. Geol. Surv. Water Resour. Invest. Rep. 89-4105, 150 pp.

Hurst, C. J., W. H. Benton, and R. E. Stetler. 1989. Detecting viruses in water. J. Am. Water Works Assoc. 81(9):71-80.

James M. Montgomery Consulting Engineers, Inc. 1979. Pilot Filtration Studies at the Tapia Water Reclamation Facility. Report prepared for the Las Virgenes Municipal Water District by James M. Montgomery Consulting Engineers, Inc., Pasadena, California.

Jensen, J. N., J. J. St. Aubin, R. F. Christman, and J. D. Johnson. 1985. Characterization of the reaction between monochloramine and isolated aquatic fulvic acid. Vol. 6, Chapter 73, pp. 939-949 in Water Chlorination: Chemistry, Environmental Impact and Health Effects, R. L. Jolly ed. Chelsea, MI: Lewis Publishers.

Keys, J. W. 1981. Grand Valley irrigation return flow case study. J. Irrig. Drain. Eng. 107(IR2):221-232.

Krasner, S. W., M. J. McGuire, J. G. Jacalangelo, N. L. Patania, K. M. Reagen, and E. M. Aieta. 1989. The occurrence of disinfection by-products in U.S. Drinking Water. J. Am. Water Works Assoc. 81:41-53.

Lambing, J. H., W. E. Johnes, and J. W. Sutphin. 1988. Reconnaissance investigations of water quality, bottom sediment, and biota associated with irrigation drainage in Bowdoin National Wildlife Refuge and adjacent areas of the Milk River basin, northeastern Montana, 1986-87. U.S. Geol. Surv. Water Resour. Invest. Rep. 87-4243, 71 pp.

Lance, J. C., R. C. Rice, and R. G. Gilbert. 1980. Renovation of sewage water by soil columns flooded with primary effluent. J. Water Pollut. Contr. Fed. 52(2):381-388.

Lang, R., B. Weinschrott, and G. Tchobanoglous. 1986. Evaluation of the Aqua-Aerobic Automatic Backwash Filter for Wastewater Reclamation in California. University of California at Davis, Department of Civil Engineering. Davis, California.

Law, J. P., Jr. 1987. Irrigation effects in Oklahoma and Texas. J. Irrig. Drain. Eng. 113(1):49-56.

Lee, E. W. 1990. Drainage water treatment and disposal options. Pp. 450-468 in Agricultural Salinity Assessment and Management, K. K. Tanji, ed. New York: American Society of Civil Engineers.

Madison, F., J. Arts, S. Berkowitz, E. Salmon, and B. Hagman. 1979. Washington County Project. EPA 905/9-80-003. U.S. Environmental Protection Agency. Chicago, Ill.

Mara, D., and S. Cairncross. 1989. Guidelines for the Safe Use of Wastewater and Excreta in Agriculture and Aquaculture: Measures for Public Health Protection. Geneva, Switzerland: World Health Organization.

Matsumoto, M. R., and G. Tchobanoglous. 1981. Evaluation of the Pulsed-Bed Filter for Wastewater Reclamation in California. University of California at Davis, Department of Civil Engineering, Davis, California.

McCarty, P. L., B. E. Rittman, and E. J. Bouwer. 1984. Microbiological Processes Affecting Chemical Transformations in Groundwater. Pp. 89-116 in Groundwater Pollution Microbiology, G. Bitton and C.P. Gerba eds. New York: John Wiley.

McComas, S. R., and D. F. Sefton. 1985. Comparison of stormwater runoff impacts on sedimentation and sediment trace metals for two urban impoundments. In Lake and Reservoir Management: Practical Applications, proceedings of the Fourth Annual Conference and International Symposium, Oct. 1984. NcAfee, N.J.: North American Lake Management Society.

Melnick, J. C., C. P. Gerba, and C. Wallis. 1978. Viruses in water. Bull. WHO 56:499-508.

Metcalf & Eddy, Inc. 1991. Wastewater Engineering: Treatment, Disposal, and Reuse. New York: McGraw-Hill.

Mossbarger, W. A., Jr., and R. W. Yost. 1989. Effects of irrigated agriculture on ground-water quality in corn belt and lake states. J. Irrig. Drain. Eng. 115(5):773-790.

Noss, C. I., R. P. Camolan, and L. Stark. 1989. Virus Removal by Wastewater Effluent Filtration. Prepared for the Florida Department of Environmental Regulation. Tallahassee, Fla.

Olivieri, V. P., C. W. Kruse, K. Kawata, and J. E. Smith. 1977. Microorganisms in Urban Stormwater. EPA-600/2-77-087. U.S. Environmental Protection Agency, Washington, D.C.

Pierce, R. C., and M. P. Wong. 1988. Pesticides in agricultural waters: The role of water quality guidelines. J. Canadian Water Resour. Assoc. (Winnipeg) 13(3):33-49.

Pitt, R. 1979. Demonstration of Nonpoint Pollution Abatement Through Improved Street Cleaning Practices. EPA-600/2-79-161. U.S. Environmental Protection Agency. Cincinnati, Ohio.

Pitt, R. 1983. Urban Bacteria Sources and Control by Street Cleaning in the Lower Rideau River Watershed, Ottawa, Ontario. Rideau River Stormwater Management Study. Ontario Ministry of the Environment. Ottawa, Ont.

Pitt, R. 1984. Characterization, Sources, and Control of Urban Runoff by Street and Sewerage Cleaning. Contract No. R-80597012. Office of Research and Development, U.S. Environmental Protection Agency, Cincinnati, Ohio.

Pitt, R., and J. McLean. 1986. Toronto Area Watershed Management Strategy Study.Humber River Pilot Watershed Project. Ontario Ministry of the Environment. Toronto, Ont.

Pitt, R., and G. Shawley. 1981. A Demonstration of Non-Point Source Pollution Management on Castro Valley Creek. Alameda County Flood Control and Water Conservation District (Hayward, CA) for the Nationwide Urban Runoff Program. Water Planning Division, U.S. Environmental Protection Agency, Washington, D.C.

Pitt, R., M. Lalor, R. Field, D. D. Adrian, and D. Barbe. 1993. Investigation of Inappropriate Pollutant Entries into Storm Drainage Systems. A User's Guide. EPA/600/- R-92/238. U.S. Environmental Protection Agency. Cincinnati, Ohio.

Pitt, R., R. Field, M. Lalor, and M. Brown. 1994. Urban stormwater toxic pollutants assessment: Sources and treatability. Accepted for publication in Water Environ. Res. To be published in late 1994.

Power, J. F., and J. S. Schepers. 1989. Nitrate contamination of groundwater in North America. Agr. Ecosyst. Environ. 26:165-187.

Qureshi, A. A., and B. J. Dutka. 1979. Microbiological studies on the quality of urban stormwater runoff in southern Ontario, Canada. Water Res. 13:977-985.

Rice, R.C. and H. Bouwer. 1984. Soil-Aquifer Treatment Using Primary Effluent. J. Water Pollut. Contr. Fed. 56(1):84-88.

Ritter, W. F., F. J. Humenik, and R. W. Skaggs. 1989. Irrigated agriculture and water quality in east. J. Irrig. Drain. Eng. 115(5):807-820.

Ritter, W. F., R. W. Scarborough, and A. E. M. Chirnside. 1991. Nitrate leaching under irrigation on Coastal Plain soil. J. Irrig. Drain. Eng. 117(4):490-502.

Rose, J. B., and R. P. Carnahan. 1992. Pathogen Removal by Full-Scale Wastewater Treatment. Prepared for the Florida Department of Environmental Regulation, Tallahassee, Fla.

Sabol, G. W., H. Bouwer, and P. J. Wierenga. 1987. Irrigation effects in Arizona and New Mexico, J. Irrig. Drain. Eng. 113(1):710-748.

Sagik, B. P., B. E. Moore, and C. A. Sorber. 1978. Infectious disease potential of land application of wastewater. Pp. 35-46 in State of Knowledge in Land Treatment of Wastewater, Vol. 1. Proceedings of an International Symposium, U.S. Army Corps of Engineers, Cold Regions Research and Engineering Laboratory, Hanover, New Hampshire.

Sanitation Districts of Los Angeles County. 1977. Pomona Virus Study: Final Report. Prepared for the California State Water Resources Control Board. Sacramento, Calif.

Schillinger, J. E., and D. G. Stuart. 1978. Quantification of Non-Point Water Pollutants from Logging, Cattle Grazing, Mining, and Subdivision Activities. NTIS No. PB 80-174063. Springfield, Va.: National Technical Information Service.

Schmidt, K. D., and I. Sherman. 1987. Effect of irrigation of groundwater quality in California. J. Irrig. Drain. Eng. 113(1):16-29.

Sheikh, B., R. P. Cort, W. R. Kirkpatrick, R. S. Jaques, and T. Asano. 1990. Monterey Wastewater Reclamation Study for Agriculture. Res. J. Water Pollut. Cont. Fed. 62(3):216-226.

Shukairy, H. M., R. J. Miller, and R. S. Summers. 1994. Bromide's effect on DBP formation, speciation, and control: part 1, ozonation. J. Am. Water Works Assoc. 86(6):72-87.

Shuval, H. I., A. Adin, B. Fattal, E. Rawitz, and P. Yekutiel. 1986. Wastewater Irrigation in Developing Countries—Health Effects and Technical Solutions. Techn. Paper No. 51. The World Bank. Washington, D.C.

Smith, W. G. 1982. Water quality enhancement through stormwater detention. In Proceedings of the Conference on Stormwater Detention Facilities, Planning, Design, Operation, and Maintenance, Henniker, New Hampshire, W. DeGroot, ed. New York: American Society of Civil Engineers.

Snider, D. E., J. L. Darby, and G. Tchobanoglous. 1991. Evaluation of Ultraviolet Disinfection for Wastewater Reuse Applications in California. Department of Civil Engineering. Davis, Calif.: University of California at Davis.

Solley, W. B., C. F. Merk, and R. R. Pierce. 1988. Estimated use of water in the United States in 1985. U.S. Geol. Surv. Circ. 1004, 82 pp.

Sonnen, M. B., J. L. Thomas, and J. C. Guitjens. 1987. Irrigation effects in six western states. J. Irrig. Drain. Eng. 113(1):57-68.

Spencer, W. F., M. M. Cliath, J. W. Blair, and R. A. LeMert. 1985. Transport of pesticides from irrigated fields in surface runoff and tile drain waters. U.S. Dept. Agr. Conserv. Res. Rep. 31, 76 pp.

Tanji, K. K. 1981. California irrigation return flow case studies. J. Irrig. Drain. Eng. 107(IR2):209-237.

Treweek, G. P. 1985. Pretreatment processes for groundwater recharge in Artificial Recharge of Groundwater, T. Asano, ed. Boston, Mass.: Butterworth.

U.S. Environmental Protection Agency. 1983. Results of the Nationwide Urban Runoff Program. Water Planning Division, PB 84-185552, U.S. Environmental Protection Agency. Washington, D.C.

U.S. Environmental Protection Agency. 1986a. Design Manual—Municipal Wastewater Disinfection, Ultraviolent radiation, Chapter 7. EPA/625/1-86/021. Water Engineering Research Laboratory, U.S. Environmental Protection Agency. Cincinnati, Ohio.

U.S. Environmental Protection Agency. 1986b. Quality Criteria for Water. EPA 440/5-86-001. U.S. Environmental Protection Agency. Washington, D.C.

U.S. Environmental Protection Agency. 1992. Guidelines for Water Reuse. EPA/625/R-92/DO4, Center for Environmental Research Information, U.S. Environmental Protection Agency. Cincinnati, Ohio.

Water Pollution Control Federation. 1989. Water Reuse (second edition)—Manual of Practice. Water Pollution Control Federation, Alexandria, Va.

Weinschrott, B., and G. Tchobanoglous. 1986. Evaluation of the Parkson DynaSand Filter for Wastewater Reclamation in California. Department of Civil Engineering, University of California at Davis, Davis, California.

Wesner, G. M. 1991. Annual Report: Orange County Water District Wastewater Reclamation and Recharge Project, Calendar Year 1990. Prepared for Orange Country Water District, Fountain Valley, California.

Williams, L. H. 1982. Effectiveness of stormwater detention. In proceedings of the Conference on Stormwater Detention Facilities, Planning, Design, Operation, and Maintenance, Henniker, New Hampshire, W. DeGroot, ed. New York: American Society of Civil Engineers.

Wilson, B., P. Roessler, E. Van Dellen, M. Abbaszadegan, and C. P. Gerba. 1992. Coliphage MS-2 as a UV water disinfection efficacy test surrogate for bacterial and viral pathogens. Poster presentation at the American Water Works Association 1992 Water Quality Technology Conference, Nov. 16-17.

3

Soil and Aquifer Processes

The desired role of the unsaturated soil zone (vadose zone) in a recharge system is a straightforward one: remove or reduce chemical and biological constituents that pose a potential health risk before the recharge water enters the ground water. Unfortunately, the processes by which removal occurs are not completely efficient in a natural setting, and not all constituents are retained or degraded to the same extent. Moreover, management strategies that may enhance the removal of one chemical or pathogen may actually decrease the efficiency of removal of another.

This chapter describes the major processes by which soils and aquifers can remove chemicals and pathogens. The processes that occur in soils and aquifers are chemical-, pathogen-, and soil-specific, depending on a number of conditions that can vary significantly from site to site and from one compound to another. Thus this chapter reviews the principal processes governing transport and fate in soil first in a generic fashion, and later with respect to the behavior of specific chemical or pathogen groups. The soil properties that are important in a properly functioning soil-aquifer treatment (SAT) system are reviewed, and properties or processes that can create difficulties for chemical and pathogen removal processes are identified. Three separate issues of concern are addressed: (1) the overall effectiveness of the SAT system and its ability to ensure that the quality of the underlying water resources will not be impaired, (2) the long-term sustainability of the system, and (3) the feasibility of monitoring to determine both the performance and the safety of the operation.

CONDITIONS INFLUENCING PRETREATMENT

The soil and aquifer properties, recharge method, type of wastewater, and ultimate destination and intended use of the recovered ground water collectively dictate the degree of pretreatment required before recharge. In addition, recharge rates, regardless of the method used, depend to some degree on the quality of the source water that is recharged. Cost-effective recharge operations are achieved through tradeoffs between maximizing recharge rates and minimizing treatment costs.

The selection of a wastewater treatment process depends on the characteristics of the wastewater, the required effluent quality, and the cost of the selected treatment option. When considering a water source as a potential for recharge, the likely variety and concentration of contaminants also need to be considered. The use of waters of impaired quality as sources for recharge has raised questions about the level of pretreatment necessary prior to recharge. The most conservative approach is to assume that passage through the soil to the aquifer and through the aquifer to the withdrawal location provides no treatment and that pretreatment processes therefore should improve the source water to the quality level needed by the end user. This approach, however, can lead to expensive systems. Soil-aquifer processes can be counted on to provide treatment benefits.

Soil Properties

The ideal porous medium for an SAT operation is one that allows rapid infiltration and complete removal of all constituents of concern. Unfortunately, no such medium exists because the attributes required to achieve one goal hamper the achievement of the other. In surface soil, coarse-textured materials are desirable for infiltration because they transmit water readily; however, the large pores in these soils are inefficient at filtering out contaminants, and the solid surfaces adjacent to the main flow paths are relatively nonreactive. In contrast, fine-textured soils are efficient at contaminant adsorption and filtration, but they have low permeability and their small pores clog easily. Structured soils containing biological channels (e.g. worm holes or root holes) or cracks are permeable, but the large flow paths completely dominate the movement of material and much of the matrix is bypassed. The best choice for an SAT soil is therefore a compromise, such as a fine sand or a sandy loam with relatively little structure (Bouwer, 1985).

Nature of Recharge Operation

Of the two basic methods of artificial recharge—surface infiltration and well recharge—well recharge requires water of much higher quality. This is particularly true where an aquifer composed of granular rocks is to be recharged.

The flux per unit area of rock surface at the point of recharge is generally much greater for injection wells than for surface infiltration systems. Consequently, equivalent amounts of clogging material and nutrients for biological growth result in more severe operational problems in recharge wells. Surface infiltration systems can function effectively over a broad range of water quality and can readily tolerate variations in quality, although, in general, higher quality results in higher infiltration rates. Injection well operations, on the other hand, are much more sensitive to quality variations, except in the case of conduit-flow rocks such as solution-riddled limestone or fractured rock.

Recharge wells require water that is virtually free of suspended matter, especially where granular aquifers are to be used. Any injected suspended matter accumulates at and near the well-aquifer interface, and because this circumferential area is limited and the flux through it is large, rapid hydraulic head loss and reduction in injection capacity occur quickly. The clogging caused by the accumulation of the suspended material and biological growth must be remedied by pumping the well for backflushing, by surging or jetting, or at times by dosing the well with chemicals to loosen and/or dissolve the accumulated clogging materials. Removal of suspended solids (to very low levels, i.e., less than 1 milligram per liter) in the recharge water is required for successful operation of recharge wells, except where karstic or fractured rock aquifers are to be recharged. Wells recharging solution-riddled or fractured rock aquifers can tolerate water having higher levels of suspended solids without experiencing severe operational problems.

Wastewater Composition

Quality parameters of concern in the operation of surface infiltration systems are the suspended solids (SS) and total dissolved solids (TDS) content as well as the concentrations of nutrients that stimulate biological growth and of major cations such as calcium, magnesium, and sodium, which determine the sodium adsorption ratio (SAR).

The suspended solids concentration of the recharge water is the most important factor. Suspended solids settle out or are filtered from the water and accumulate on the soil and/or at a short distance below the water-soil interface. The accumulation reduces the permeability of the soil and retards movement of water into the subsurface.

Swelling and deflocculation of clay minerals contained in the aquifer can occur if the recharge water contains a higher ratio of monovalent to divalent cations (higher SAR) than does the native water. Clay dispersal can also occur if freshwater is recharged into a saline aquifer. The combination of low TDS with high SAR causes clays to disperse. If clays disperse, infiltration rates drop. Chemical reactions between the recharge water and the native ground water can cause precipitates to form and these can clog pores and reduce injection capacity. Compounds of carbonates, phosphates, or iron oxides are the most likely to cause such problems. To avoid problems caused by clay dispersal or chemical precipitates, an evaluation of the chemical compatibility of the recharge water with the aquifer materials and the native ground water should be conducted. The evaluation may indicate the need to chemically modify the recharge water to avoid clogging problems caused by chemical incompatibility.

Water recharged by wells must be free of entrained or dissolved gases that may evolve when cold recharge water is injected into warmer ground water. Entrained air or gases that come out of solution will reduce the aquifer's hydraulic conductivity and, consequently, the injection capacity of the well.

Bacterial growth is yet another concern with regard to clogging of recharge wells. Removal of nutrients and biodegradable matter from the recharge well and disinfection of the water minimize the potential for bacterial growth in the immediate vicinity of the recharge well. When the recharge water contains a chlorine residual, well clogging is slower (Vecchioli et al., 1980). An alternative to continuous chlorination of the recharge water is to backpump the well frequently (about once a day) or to dose the well heavily with chlorine periodically to destroy the bacterial growth and then backpump the well to remove the spent chlorine solution and organic residue.

Recharge wells ending in the vadose zone (dry wells) cannot be redeveloped readily because it is not possible to pump water from them to remove accumulated suspended matter or other clogging materials. Therefore, water recharged to dry wells must be free of suspended solids and not cause clay dispersal or bacterial growth. On the other hand, because of the generally shallow depths of dry wells, replacing clogged wells is considerably less costly than for wells injecting water below the water table.

GENERAL DESCRIPTION OF SUBSURFACE PROCESSES

Vadose Zone Processes

The ultimate goal of a ground water recharge project is to resupply the subsurface with water that does not impair the quality of the underlying resource. Thus, the role of the unsaturated or vadose zone in recharge systems is to help filter out or transform harmful constituents in the soil solution as recharge water moves through the soil matrix en route to the aquifer.

The vadose zone is a much more complex transport medium than an aquifer, for several reasons. Because only part of the void space is filled with water, chemicals with a significant vapor pressure can move in the gas phase as well as in solution. The water flow rate can vary significantly. The resistance offered by the vadose zone to the flow of water through a given local soil volume is a nonlinear function of the water content, whereas in the saturated zone, it is a constant. The temperature varies in the surface regime in response to the cyclic inputs of radiant energy, and the composition of the air, solid, and solution phases of the soil is also dynamic, causing spatial and temporal variations in the chemical and biological reactions that transform chemicals in the vadose zone. Also, the amount of water retained against gravity varies significantly with soil texture. Coarse-textured, sandy soils may hold as little as 10 to 20 percent of water-saturation after drainage becomes insignificant, while fine-textured silts or clays may hold as much as 90 percent. Restricting layers comprised of clay lenses or cementing agents can retard drainage greatly, even in otherwise permeable media. Soils that retain extensive water are prone to aeration problems.

There are many different processes that can remove chemicals or pathogens from the recharge water as it flows through the vadose zone. Some chemicals volatilize and escape to the atmosphere. They can be chemically or biologically transformed to a new form that may or may not be toxic. They can attach to stationary soil mineral or organic surfaces or precipitate out of solution. They can form complexes with dissolved constituents or particulate matter in solution, thereby reducing their attraction to the soil solid phase and enhancing their mobility in solution. Large pathogens such as parasites, some bacteria, and colloidal material containing contaminants can be filtered out of solution by narrow soil pores, a process that slowly clogs the medium and eventually reduces its permeability if the contaminants are not biodegraded. Viruses can be retained by soil solid phases and inactivated by reactions occurring in the soil.

Aquifer Processes

Chemicals or pathogens that are still present in solution when the recharge water reaches the aquifer are subject to many of the same processes that occur in the vadose zone, with several exceptions. The biological activity in ground water is much slower than in the near-surface zone, so degradation is greatly reduced. Water fills all of the pore spaces within the ground water zone, so the only place where volatilization can occur is in the capillary fringe above the water table interface. Aquifers used for ground water recharge projects are generally much more coarse textured than soils, so colloids and large pathogens, should they still be present in ground water, are not as easily filtered out of solution as they are in surface soil.

Volatilization

Volatilization refers to the evaporation of chemical vapor from soil or water bodies and its subsequent loss to the atmosphere. Many organic compounds are volatile in water, as are some nitrogen compounds (e.g., ammonia, nitrous oxide) generated from biological transformations. In addition, some inorganic chemicals (e.g., selenium compounds) may be rendered volatile through biological reactions.

The chemical characteristic that is most indicative of a compound's volatility is its Henry's constant, which is the ratio between the vapor pressure and the solubility of the pure chemical in water. It may also be expressed in dimensionless form as the ratio between the chemical concentration in the gaseous and aqueous phases,

$$K_H = \frac{C_g}{C_w} \qquad (1)$$

where C_g is the mass of chemical vapor per unit volume of air and C_w is the mass of chemical dissolved per unit volume of aqueous solution.

For a given chemical, the extent of volatilization loss is very dependent on the soil and atmospheric conditions. The primary factor determining loss is the air phase concentration maintained at the interface with the atmosphere. In general, volatilization is greatly reduced in soil compared to water because the soil solid phase retains the chemical mass, thereby reducing its vapor pressure. In addition, the soil can offer substantial resistance to the transport of chemical from the soil profile to the surface, particularly if the soil is wet and little upward flow of water is occurring. Therefore, in a typical SAT process where recharge water is ponded over the surface for prolonged periods of time, the primary route for volatilization loss will be from the surface of the standing water. During the drainage cycle when the soil becomes unsaturated, volatile constituents in solution near the surface can also evaporate and escape to the atmosphere.

Volatilization from Surface Water

Volatilization from standing water can be represented conceptually as a two-film resistance model, in which the dissolved compound moves from the bulk fluid through a liquid film to the evaporating surface, and then diffuses through a stagnant air film to the well mixed atmosphere above (Liss and Slater, 1974). The two-film model assumes that the chemical is well mixed in the bulk solution below the liquid film and that mass transfer across each film is proportional to the concentration difference. With these assumptions, a chemical in the water body volatilizes at a rate proportional to the bulk concentration, so that the entire loss process can be characterized by an effective "half-life," defining the amount of time required to reduce the mass in solution by 50 percent. Thomas (1982) reviewed volatilization loss models for chemicals present in water bodies and performed model calculations for a number of compounds. The film thicknesses depend on specific conditions within the water and air, but may be crudely estimated from default values given in Thomas (1982) when no actual data are available. Table 3.1 summarizes effective volatilization half lives calculated for a range of Henry's constant values.

As seen from Table 3.1, the effective half-life varies widely, depending on

TABLE 3.1 Effective Volatilization Half-Life Ranges as a Function of Dimensionless Henry's Constant Values for a Stagnant Water Body of 1-m Depth

Henry's Constant K_H	Example	Half life (days)
10^{-8} - 10^{-7}	Bromacil	10^4 - 10^5
10^{-7} - 10^{-6}	Atrazine	10^3 - 10^4
10^{-6} - 10^{-5}	Phenol	10^2 - 10^3
10^{-5} - 10^{-4}	Diazinon	10 - 100
10^{-4} - 10^{-3}	EPTC	1 - 10
10^{-3} - 10^{-2}	Bromobenzene	~ 1
10^{-2} - 10^{-1}	Benzene	~ 1
10^{-1} - 1	Methyl bromide	~ 1
1 - ∞	Vinyl chloride	~ 1

Source: Default values for film transfer coefficients are taken from Thomas (1982), and K_H values are taken from Jury et al. (1984b).

the value of the Henry's constant. Clearly, compounds with half lives that are considerably less than the detention time of the water on the surface will not enter the soil. For instance, volatilization losses of 22 to 73 percent were found for a wide spectrum of hydrocarbons in sewage effluent infiltration basins in Phoenix, Arizona (Bouwer et al., 1986).

Volatilization from Soil

The volatilization loss rates of chemicals from soil are generally smaller than those from standing water for several reasons. First, the diffusion resistance of soil is greater than that of free air because of the solid and liquid barriers to gas movement. Second, adsorption of chemical to soil solids reduces the vapor pressure of the compound by removing mass from solution. Because the transport pathways from the soil to the surface are much more complex than in free water, the two-film resistance model of volatilization is not applicable to soil, and more sophisticated estimation methods must be used.

Jury et al. (1983, 1984a,b,c, 1990) developed a comprehensive screening model for evaluating volatilization losses of chemicals after their incorporation into a soil layer of arbitrary thickness. They performed calculations covering a range of initial conditions on a large group of organic compounds, allowing the volatilization losses of different chemicals to be compared and grouped. In general, they found that compounds with dimensionless Henry's constant values less than 10^{-4} were not prone to significant volatilization after deposition in the soil, but that extremely volatile compounds could move upward to the surface from substantial distances if they were not rapidly degraded. Under certain conditions, water evaporation can greatly enhance volatilization from soil by concentrating the chemical mass at the surface and raising its vapor pressure.

There are several good references (see Howard, 1990; Howard et al., 1991) containing compendia of chemical properties and soil-chemical interaction coefficients (such as sorption coefficients and degradation rate coefficients). These are more reliable sources of information on chemical properties than earlier references, which frequently contain outdated or inaccurate data.

Volatilization During Ground Water Recharge

Chemicals likely to volatilize during ground water recharge operations include nitrogen compounds and organics with high Henry's constant values. Operations conducted under high water saturation, such as ponding, will minimize the loss of organics from soil by volatilization because these will be blocked from entering the gas phase and reaching the soil surface if the soil air space is filled with water. However, the presence of standing water on the surface enhances the loss from water, so that any chemicals with volatilization half-lives that are significantly less than the total detention time in surface water will probably evaporate out of solution before they ever reach the soil.

Recharge water containing fertilizer or sewage contributions may contain nitrogen compounds that can be transformed to volatile species. Ammonia is very volatile and vaporizes from anhydrous form immediately upon exposure to the air. The ammonium ion NH_4^+, which is a constituent of many fertilizers, partitions into a volatile gas when exposed to air; therefore, any dissolved NH_4^+ near the atmospheric interface will lose nitrogen to the atmosphere. NH_4^+ is readily transformed by soil bacteria to nitrate (NO_3^-), which is very mobile in soil and is nonvolatile. However, NO_3^- may be transformed anaerobically by biological denitrification to several gaseous species (primarily N_2O and N_2) when soil water content is high and a source of organic carbon is present. Ground water recharge by ponding may therefore enhance removal of residual nitrogen through this process.

There is evidence that potentially harmful inorganic compounds of selenium and other trace metals may be removed from contaminated soil by volatilization after the compounds have been methylated through biological transformation (Karlson and Frankenberger, 1989).

Transport of Dissolved Chemicals

Unless a compound is very volatile, it moves primarily in the aqueous solution phase. At the scale of the soil pore, there are two transport mechanisms that can move solutes through the medium: convection and diffusion. Convection is the transport of a dissolved chemical by virtue of bulk movement of the host water phase, while diffusion is the random mixing caused by collisions at the molecular scale. The local water flux describes three-dimensional flow around the solid and gaseous portions of the medium and is not measurable. Instead, the local quantities are volume averaged to produce a larger-scale representation of the system properties. The averaging volume must be large enough that the statistical distribution of geometric obstacles is the same from place to place. If the porous medium contains the same material and density throughout, then the mean value produced by this averaging is macroscopically homogeneous over the new transport volume containing the averaged elements. The large-scale convective solute flux is then the product of the average water flux and the average solution concentration C_w.

However, in the process of averaging, some of the solute motion is lost. The small-scale migration of solution through tortuous pathways within the porous medium no longer appears in the water flux expression after volume averaging, because the fluctuations about the mean motion do not contribute to the net water transport. Dissolved chemicals convected along these tortuous flow paths do contribute to solute transport and cannot be neglected. The motion of solute due to small-scale convective fluctuations about the mean motion is called mechanical dispersion (Bear, 1972). The combined mixing associated with diffusion and mechanical dispersion is called hydrodynamic dispersion. Development of models to describe the hydrodynamic dispersion flux is one of the most active research areas in soil physics and hydrology today (Dagan, 1986). The most widely used representation of dissolved chemical movement is the convection-dispersion model, which assumes that the dispersive mixing is random or diffusionlike within the moving fluid.

Solute Velocity

The water flux (or Darcy velocity) q is the volume of water flowing per unit time per unit area of soil. Because the water flows only through the water-filled regions of the porous medium, the actual average velocity of the water is equal to

$$v = \frac{q}{\theta} \qquad (2)$$

where θ is the volumetric water content (volume of water per volume of soil).

Equation (2) describes the average linear velocity of the water and therefore also represents the movement of a chemical that travels freely with water and does not interact with solid surfaces in the soil. At normal pH levels, negatively charged ionic species such as NO_3^- are not attracted to soil mineral or organic surfaces and move relatively freely with flowing solution. In fact, the negatively charged clay lattice actually has a repulsive effect on the anions in solution, causing them to avoid the solution region that is closest to the surfaces, in which case (2) will somewhat underestimate the solute velocity. This effect is minor unless the soil is high in clay.

Therefore, a chemical moving in solution at a water flux q that flows through a water-filled volume fraction θ without interacting with the solid phase will move with a velocity given by (2) and hence will require a time

$$t = \frac{L}{v} = \frac{L\theta}{q} \tag{3}$$

to move a distance L through the soil or aquifer. Because of diffusion and dispersion, not all of the solute molecules will move at the same speed, but will spread out around the average arrival time given in (3).

Solute Dispersion During Transport

The effect of dispersion is to produce chemical spreading during transport. It is greatly affected by soil structure and depends also on the scale over which the water flux is volume averaged. Soils with a pronounced macrostructure can transmit chemicals rapidly, leading to early arrival that greatly precedes the main pulse or front. In addition, aggregated regions of the medium containing water that is not flowing can act as repositories for diffusing chemicals, causing portions of a front or pulse to lag far behind the average flow.

Dispersion modeling is poorly developed in unsaturated field soils. Near the soil surface of an alluvial soil, it is likely that dispersion will be dominated by differences in water velocity at different points in the soil. In this case, the common convection-dispersion model assuming random solute spreading is not likely to be accurate, and a stochastic convective model assuming parallel flow of solute in isolated stream tubes may represent the solute mixing process better (Butters and Jury, 1989).

Solute Sorption During Transport

Chemicals that are hydrophobic or positively charged do not travel at the speed of the flowing water, but rather are slowed by their attraction to stationary solid sorption sites. Although the sorption process is very dynamic at the molecular scale, it is useful to conceptualize an adsorbed molecule as existing in a distinct phase that is temporarily immobilized by virtue of its attachment to stationary solid matter.

There are several different types of sorption reactions, distinguished primarily by the nature of the sorbing surface and the charge characteristics of the sorbing molecule. Positively charged ions in solution are attracted to negatively charged clay mineral surfaces and are temporarily immobilized by the process known as cation exchange. This is a partitioning reaction that divides the chemical mass between solution and adsorbed phases; it does not completely strip a compound from solution, nor is it permanent. The reaction depends on the nature of the molecule and also on the composition of the soil solution (Sposito, 1981). In addition, some positively charged species, notably the trace metals, appear to be specifically sorbed strongly to certain oxide surfaces (Chang and Page, 1985).

Anions are repelled from clay mineral surfaces that are negatively charged, but are attracted to positively charged broken end faces of minerals and also to free oxides in the soil. These surfaces have charges that are strongly pH dependent, and attract anions most strongly under acidic conditions.

Neutral organic molecules such as nonionic pesticides sorb primarily to organic matter surfaces in a reaction that can be approximated by a partition coefficient. The form of this relation is denoted through the linear sorption model

$$C_s = K_d C_w \tag{4}$$

where C_s is the sorbed chemical concentration (mass of chemical sorbed per mass of soil), C_w is the chemical concentration in the soil solution (mass of chemical per mass of water), and the proportionality coefficient K_d is called the distribution coefficient (Hamaker and Thompson, 1972). The distribution coefficient has units of volume per mass. Because most of the sorption occurs on organic matter surfaces, the distribution coefficient may be subdivided into

$$K_d = f_{oc} K_{oc} \tag{5}$$

where f_{oc} is the fraction of soil organic carbon content (mass per mass of soil) and K_{oc} is the distribution coefficient per unit organic carbon, called the organic carbon partition coefficient. Hamaker and Thompson (1972) showed that the K_{oc} of a given compound varied significantly less between soils than the K_d. In that sense, it represents an intrinsic sorption affinity for a chemical that is soil independent. The organic carbon partition coefficient for chemicals generally decreases with increasing water solubility.

The retardation factor R is another measure of the relative partitioning of a chemical between the soil and the aqueous phases:

$$R = 1 + \frac{\rho_b K_d}{\theta} = 1 + \frac{\rho_b f_{oc} K_{oc}}{\theta} \tag{6}$$

where ρ_b is the soil bulk density. R is the ratio between the total mass density and the mass per soil volume θC_w in the dissolved phase. A simple derivation of equation (6) was presented by Bouwer (1991). Because this relation applies instantaneously at equilibrium, it may be interpreted in probabilistic terms as describing that the sorbing solute spends $1/R$ as much time in the dissolved phase as an identical solute that does not sorb; hence, it will take R times as long to travel a given distance as the mobile one. Therefore, the solute velocity V_R of an sorbing chemical with a linear sorption isotherm is equal to

$$V_R = \frac{q}{\theta R} = \frac{v}{R} \qquad (7)$$

Thus, a sorbing molecule moves R times slower than a nonsorbing one under the same conditions.

Equations (2) and (7) may be used to predict average migration of mobile and sorbing chemicals, provided that the assumptions of complete mobility of the water in soil and linear, equilibrium sorption of the chemical are met. Table 3.2 shows travel times required for different chemicals to move 50 meters (m) (160 ft) in ground water under the conditions given. The predictions apply only to the center of the pulse or front and do not address the extreme movement of the leading or trailing edges. The retardation factors in this table show a variability of nearly 4 orders of magnitude in the solute velocity and in the retention times for chemicals in the subsurface environment.

Although the travel time model describing average movement is simple, it is useful. Pratt et al. (1972) applied it to interpret deep cores taken underneath citrus groves throughout California. They found nitrate fronts moving very slowly underneath these fields and estimated that as much as 50 years would be required for the chemicals to reach deep ground water aquifers under the fields. Pesticides or other contaminants that also adsorb to soil solids would have considerably longer travel times under the same conditions. In contrast, SAT systems with relatively high flow rates and smaller distances from the soil surface to ground water may have much smaller travel times. For example, an SAT system recharging 50 m (160 ft) of water per year to ground water at a depth of 4 m (13 ft) through a soil layer with the same properties as the model aquifer in Table 3.2 would produce a travel time of only 11.7 days for mobile chemicals such as NO_3^-.

Other Attenuation Mechanisms

Chemical and biological reactions occurring during passage through an SAT system are specific to the compounds moving through them and are discussed in detail in the next section. Pathogens such as helminth cysts, protozoa, and bacteria are large enough to be filtered by the smaller soil pores and can be permanently immobilized near the surface, where they die and decompose, unless the

TABLE 3-2 Travel Times Required to Move 50 Meters Calculated for Different Chemicals in Ground Water.

	K_{oc} (m³/kg)	R	Travel Time (years)
Atrazine	1.6×10^{-1}	5.8	2.3
Benzene	8.3×10^{-2}	3.5	1.4
Bromide	≈ 0	1.0	0.4
DBCP	7.0×10^{-2}	3.1	1.2
DDT	$2.4 \times 10^{+2}$	7,200	2,900
Ethylene dibromide	4.4×10^{-2}	2.3	0.9
Toluene	9.8×10^{-2}	3.9	1.6
Trichlororethylene	1.4×10^{-1}	5.2	2.1
Trifluralin	7.3	220	88
Vinyl chloride	4.0×10^{-1}	13	5.2

[a]Saturated Water Content $\theta = 0.4$; bulk density $\rho_b = 1500$ Kilograms per Cubic Meter; and Organic Carbon Fraction $f_{oc} = 0.008$, which is flowing at a Rate $q = 50$ meters per year.

Source: Jury et al. (1984c, 1990).

soil is very coarse textured. Colloidal complexes formed with mineral or organic matter suspended in solution are subject to the same filtration action.

IMPORTANT SOIL-AQUIFER PROPERTIES

Vadose Zone Properties

The effect of the soil properties in ground water recharge depends on the method of recharge used. For unconfined aquifers recharged by surface application to the soil above in an SAT system, knowledge of the physical, chemical, and biological properties of the soil materials is critical to the design of the operation. In contrast, for confined aquifers, where reclaimed water is injected directly into the ground water strata, the properties of the overlying soil layer are irrelevant to the operation.

The most critical soil properties for surface application are texture, permeability, presence of clay, iron, or hardpan, depth of soil profile, presence of organic matter, and compaction characteristics (American Society of Civil Engineering, 1961). For a spreading basin operation, a number of factors must be considered during evaluation of a potential facility. Foremost among these are the quantity of recharge water available for application to the facility, the rate at which the recharge water will be applied, the quality of the recharge water, and available subsurface storage (Asano, 1985).

When this information has been assembled, the suitability of the site can be assessed by evaluating its capacity for accepting recharge water and transferring it to the aquifer. At this point, the infiltration capacity and uniformity are evaluated and a verification is made that the profile is free of subsurface barriers to downward flow, free of factors promoting clogging (such as fine layers near the surface), and free of conditions that might cause cracks to form during dry periods. Basically, the potential infiltration rate for clear water (tap water) and no clogging is equal to the "rewet" or "resaturated hydraulic conductivity." The water content after drainage (the so-called field capacity) should also be relatively low for aeration purposes. Vadose zones should also be free of undesirable chemicals (anthropogenic or natural) that can be leached out. Aquifers should be free of pollution zones that could be moved to undesirable places with the recharge flow.

The vadose zone must function as a water purification system as well as a conduit to the aquifer. Depending on the operation and the constituents of the recharge water and soil, the following physical, chemical, or biological transformations of contaminants within the water may occur: filtration of suspended solids, parasites, and bacteria; sorption of trace elements, bacteria, and viruses; precipitation of phosphates and trace metals; biodegradation of organics; recarbonation of high pH effluents; and denitrification (Asano, 1985).

To assess the soil potential for water purification, certain chemical characteristics of the soil properties, such as organic matter content, pH, and cation exchange capacity (CEC), must also be characterized.

The soil properties desirable for the full range of operations are a compromise between the optimal properties required for portions of the full operation. Coarse-textured soils are highly desirable for rapid infiltration, but perform poorly in filtration and chemical transformation of undesirable constituents of the water (Nellor, 1980). Therefore, a compromise texture (sandy loam, fine sand) with sufficient clay for sorption and filtration but high enough permeability to accept high water infiltration rates is generally the best choice for an SAT system (Bouwer, 1985).

The final choices for site characteristics are dictated by the economics of the operation and availability of suitable land. The susceptibility of physical features, such as slope or terrain, to excessive runoff during intense rainstorms should also be considered.

Control of Ground Water Flow

If recharge water is to be extracted and reused, then the ground water flow must be controlled. Such control of subsurface hydraulics usually is maintained through injection or production wells. The purpose of such control is not only to segregate the recharge water from the native aquifer water, but also to achieve a minimum desired detention time for recharge water in the subsurface. For instance, the recharge system at El Paso, Texas, was designed for a residence time of 2 years; the residence time determined from ground water monitoring was actually three times longer.

If water is recharged to take advantage of the assimilative capacity of the subsurface as part of a treatment train, then it is generally desireable to minimize or control the mixing between recharge and native aquifer waters. The recharge water, once collected, may undergo further treatment before reuse. In addition, in coastal areas and in certain arid regions, recharge waters may be cycled through aquifers containing waters of poor quality, and here it is also desirable to segregate the recharge and native waters so as not to impair the quality of the recharge water prior to extraction.

To a large extent, the detention time of recharge water during subsurface flow determines the improvement in water quality. In designing recharge systems, it is desirable to maintain a minimum residence time in the subsurface as well as a variation of residence times for different parcels of recharge water. If the control system is designed to achieve a linear variation of residence times over a specified period, then arbitrary variations in quality of recharge water over the same period may be averaged out during subsurface flow (Huisman and Olsthoorn, 1983).

UNDESIRABLE SOIL CHARACTERISTICS

No field soil possesses ideal properties for all SAT operations. The extreme diversity of the soil environment ensures that there will be extremes in the local values of the hydraulic or retention properties of the medium that can cause problems in the operation or monitoring of recharge systems. The most important of these characteristics are discussed below.

Spatial Variability of Soil Properties

Soil properties, particularly those that influence transport, vary significantly over space, as shown in Table 3.3. For example, the saturated hydraulic conductivity, a reciprocal measure of the resistance of saturated soil to the flow of water, can have a coefficient of variability of several hundred percent on the scale of an agricultural field (Jury, 1985). Such variability creates enormous data requirements and also places extreme demands on model description of

TABLE 3.3 Sample Coefficient of Variation (CV) Measured for Different Soil Water and Solute Properties in Unsaturated Fields

	Number of Studies	Range of CV (%)
Porosity	4	7-11
Bulk density	8	3-26
Water content at 0.1 bar	4	4-20
Percentage sand or clay	5	3-55
pH	4	2-15
Saturated hydraulic conductivity	12	48-320
Infiltration rate	5	23-97
Unsaturated hydraulic conductivity	4	41-343
Local solute velocity		
Steady unsaturated water flow	3	36-75
Ponded water flow	2	78-194

extreme flow events. For example, in a field that has a coefficient of variability of 200 percent in its infiltration rate, nearly half of the water entering the soil will move through the 10 percent of the field having the highest infiltration rates (Jury et al., 1991). Prediction of extreme flow under conditions of such high variability often is not feasible.

Moreover, when variability of infiltration rate is high, basic calculations of the mean travel time and assimilative capacity of the soil are greatly in error. For the example described above, the 10 percent of the field receiving half of the water will have a travel time for mobile chemicals that is about one-fifth of the average and will exhaust its exchange complex in one-fifth the time calculated for the field based on average soil properties.

Modern theories of transport through media having spatially variable properties have addressed this problem by developing stochastic models, whose predictions are expressed in probabilistic rather than deterministic terms. Although stochastic modeling is highly developed for ground water flow (Dagan, 1986), little work has been done in unsaturated soils.

Spatial variability is primarily important where the recharge water is very clean and infiltration rates are controlled by soil hydraulic conductivity. Where the water contains suspended solids, organic carbon, and/or nutrients, a clogging layer will be formed on top of the soil surface. Such clogging layers have a high hydraulic impedance, and they become the controlling factor of infiltration rates, which, of course, are much lower in clogging layers than in clean soils. Clogging develops faster where infiltration rates are high. Hence, clogging may well be the great equalizer in the infiltration process, and it may well render infiltration to be much more uniformly distributed than indicated by the spatial variability of the soil.

Preferential Flow of Chemicals

Preferential flow refers to the faster-than-average movement of contaminants or water, by whatever means, through part of the soil volume. It differs from the extremes in flow caused by property heterogeneity in that the nature of the flow is different from that in the bulk soil matrix. A major cause of preferential flow is the presence of numerous geometric voids in a structured or biologically diverse soil. For example, Ritchie et al. (1972) used a visible dye to demonstrate that much of the water moving through a swelling clay soil was migrating through vertical cracks. Omoti and Wild (1979) used fluorescent dyes in a weakly structured loamy sand to determine that earthworm channels, fissures with apertures between 0.05 and 0.10 millimeters (mm) (0.002 and 0.0039 inches), and loosely packed soil were all acting as conduits for the rapid transport of the adsorbing dyes they used in their field study. Scotter and Kanchanasut (1981) reported movement of chloride to a mole drain at 0.4 m (16 inches) under continuous ponding within 5 minutes after introduction of the chemical into the infiltrating water. Dye tracing revealed that root and worm channels and occasional fracture planes were carrying the flow.

Structureless sandy soils have also been found to exhibit preferential flow, arising from a variety of causes. Kung (1990) excavated a field plot on Plainfield sand after dye application, finding that the water funneled into increasingly narrow zones at greater depths, eventually carrying the bulk of the flow in a small fraction of the cross-sectional area. He postulated that the funneling mechanism was discrete coarse sand lenses that acted as barriers to downward movement, causing the water to flow around them and focus at the edges. A similar phenomenon was reported by Ghodrati and Jury (1990), who observed that preferential flow regions constituting a small fraction of the cross section were moving more than twice as deep as the main front under both ponded and sprinkler irrigation in undisturbed field plots on a loamy sand.

Preferential flow may also originate from unstable fluid flow through the soil. This phenomenon has a number of possible causes, including density and viscosity differences between the antecedent and incoming water, air entrapment during infiltration, and downward flow from a fine-textured to a coarse-textured soil layer (Hillel, 1987). Preferential flow has serious implications for chemical and pathogen movement because it is desirable to keep these compounds from migrating below the surface zone. Based on the information from studies conducted under controlled conditions in the field, preferential flow is widespread and often significant and also has been observed to occur in the structureless fine sandy soils that are favored in SAT systems (Jury and Fluhler, 1992). However, the presence of clogging material in SAT systems may act to prevent funnel flow from occurring to any great extent in SAT because this mechanism depends on very high permeability in the portion of the matrix that is active in flow (Kung, 1990). A more likely circumstance that could cause preferential flow in an SAT system is the presence of a more permeable zone located beneath a less conducting one, which can create fingering of fluid through narrow channels within a small part of the pore space of the more coarse-textured soil (Hillel, 1987).

Reviews of the prevalence of preferential flow and its importance in soil are given elsewhere (Beven and Germann, 1982; Jury and Roth, 1990; Jury and Fluhler, 1992).

TRANSPORT AND FATE OF SPECIFIC CONSTITUENTS OF RECHARGE WATER

The principles discussed in the previous sections are useful to understanding the chemical and pathogen removal processes that occur during transport through soil and aquifer material. This section looks at categories of solution constituents that correspond to groupings used elsewhere in this report.

Nitrogen

Nitrogen is a common constituent of wastewater and agricultural return flow. In the former, it is mostly present in the form of NH_4^+ after conventional primary and secondary treatment, as it first makes contact with the soil surface. Because NH_4^+ partitions to gaseous ammonia at the air-water interface, some volatilization occurs during the detention time and later during drainage cycles that expose soil NH_4^+ remaining near the surface.

After NH_4^+ enters the soil, it is biologically transformed to NO_3^- by a two-stage process called nitrification. In contrast to NH_4^+ which is positively charged, NO_3^- is an anion and is quite mobile in soil. However, NO_3^- can be biologically transformed under conditions of high water content and sufficient organic carbon availability into nitrogen and nitrous gases, which escape into the atmosphere. This reaction occurs when nitrogen substitutes for oxygen as a terminal electron acceptor under conditions of limiting aeration, moderated by the availability of a carbon source. Since organic carbon is normally low in aquifers, denitrification is generally confined to the near-surface regime. Crites (1985)

TABLE 3.4 Nitrogen Removal Rates at SAT Sites

	Loading Rate (feet/year)	Flooding: Drying Time	BOD:N[a]	Percentage Nitrogen Removal
Hollister, California	50	1:14	5.5:1	93
Brookings, South Dakota	40	1:2	2:1	80
Calumet, Michigan	56	1:2	3.6:1	75
Phoenix, Arizona	200	9:12	1:1	65
Ft. Devons, Massachusetts	100	2:12	2.4:1	60
Lake George, New York	190	1:4	2:1	50
Disney World, Florida	180	150:14	0.3:1	12

[a]BOD:N = ratio of biochemical oxygen demand to nitrogen.

Source: Crites, 1985.

states that carbon is required at a BOD (biochemical oxygen demand) to nitrogen ratio of about 3:1 for maximum denitrification.

Under SAT conditions the removal rates of nitrogen can be quite high if denitrification is optimized. Bouwer (1985) found that flooding and drying cycles allowed NH_4^+ to adsorb to soil mineral surfaces by cation exchange, while the drying cycle permitted oxygen diffusion that promoted nitrification to the NO_3^- form, which subsequently denitrified after diffusion into anaerobic microsites. Table 3.4 shows nitrogen removal efficiencies measured at a number of SAT sites throughout the United States. The primary removal mechanism in these operations is denitrification, with perhaps some ammonia volatilization losses during drying cycles.

Phosphorus

Phosphates in recharge water are removed by precipitation to amorphous or crystalline forms with iron, aluminum, or calcium. At low pH, precipitation with iron and aluminum is favored, whereas under alkaline conditions, calcium phosphate is controlling. Phosphate mobility is greatest under neutral conditions (Bouwer, 1985).

Phosphorus removal is achieved through a fast sorption reaction and slow precipitation reactions. The final phosphorus concentration in the effluent water after passage through the SAT system is controlled by the solubility products of the solution constituents; for that reason, removal is not complete and depends on soil properties and loading rates. The duration of the flooding cycle can be varied to promote optimal levels of denitrification for nitrogen-rich recharge waters.

Inorganics and Trace Metals

The organic constituents found in municipal wastewater tend to have been added during water use, while inorganic chemical concentrations tend to be characteristic of the source water. Moreover, they are not removed to any extent during use, but rather tend to accumulate as water is extracted from the source stream (Chang and Page, 1985). For that reason, mineral concentrations in municipal wastewater vary widely depending on the genesis of the water and its use prior to arrival at the SAT system. In contrast, inorganic chemical increases, especially in heavy metals, are more typical of stormwater runoff.

Trace elements present in suspended matter generally are removed during SAT by filtration and do not migrate. However, they do accumulate in colloidal material trapped in the clogging layer, which eventually must be removed from soil to restore its infiltration rate. Smaller suspended particulates that can move through soil pores without becoming trapped are also attenuated by sorption to mineral surfaces in the soil matrix. Chang and Page (1985) state that filtration and associated colloidal sorption are the primary means of removal of trace elements in soil.

Not all of the trace element mass is associated with suspended material in wastewater; it is also present in a dissolved form that is not affected by filtration. The dissolved constituents are affected by various chemical reactions in soil, some of which act to attenuate their movement through soil. Aerobic conditions and a high pH also decrease trace metal mobility.

Positively charged trace metal ions can be attenuated by ion exchange with negatively charged clay mineral surfaces. This process is a partition reaction affecting all cations in soil and therefore is not a significant removal mechanism for trace elements at low values. Moreover, soil cation exchange capacity is finite and therefore can be exhausted after many pore volumes of material have passed through the profile.

A more specific sorption reaction appears to occur with trace elements, which are strongly attracted to amorphous and crystalline oxides of iron, aluminum, and magnesium. This reaction is not easily reversed by other ions in solution and may become nonexchangeable over time (Jenne, 1968; Sharpless et al., 1969).

Precipitation reactions of trace metals also occur with other constituents of wastewater, which causes the concentrations in solution to be regulated by the solubility products of the species present. This reaction is difficult to separate from specific sorption.

Metals can also form complexes with dissolved organic matter (chelation), increasing their mobility in the process. Because of the complexity of solution chemistry when organic ligands are present, little is known about the behavior of specific complexes.

Because trace metal removal is incomplete and the soil sorption capacity is finite, SAT is not completely effective in preventing migration of all constituents within this group. There are also significant variations in trace element mobility, ranging from strongly attenuated (zinc and copper), to more mobile (cadmium and lead), to extremely mobile (boron). Ground water concentrations of silver, barium, cadmium, cobalt, and chromium below the treatment site at Hollister, California, have been unaffected by the additions of wastewater to the overlying soil. However, manganese, nickel, iron, zinc, lead, and copper were above background levels (Crites, 1985). Soil samples taken in the Whittier Narrows recharge plant after over 20 years of operation showed elevated levels of cadmium, chromium, copper, nickel, lead, and zinc in the top 60 cm (2 ft), but not below that depth, suggesting that the soil had the capacity to remove metals for many more years of operation before ground water would be affected (Chang and Page, 1985).

Organic Chemicals

Organic chemicals vary enormously in their mobility, volatility, and persistence in soil. In an SAT system, volatile compounds volatilize prior to application, and only the soluble organic constituents enter the soil. The organic removal efficiency of an SAT system depends on the degree to which a given compound can be chemically or biologically transformed to an innocuous state during its time of passage through the system. Organic compounds degrade chemically by hydrolysis, photodecomposition, or redox reactions to varying degrees depending on their structure (Armstrong and Konrad, 1974). Microbial conversion occurs chiefly in the surface zone of soil, where bacterial populations and organic carbon levels are high. Low organic carbon levels generally limit microbial action in deeper regions of the vadose zone and in aquifers.

As discussed in a previous section, the travel time of an organic chemical may be roughly designated by its retardation factor in a given soil, or more intrinsically by its organic carbon partition coefficient K_{oc}. In a similar manner, the overall action of the chemical and microbiological processes transforming an organic chemical moving in soil may be crudely expressed as a half-life or degradation rate constant. Then, the length of the half life compared to the travel time can be used as an index of the potential for the compound to survive its passage through the soil. This approach has been used to develop pesticide screening models to determine (in a relative sense) the potential of a compound to reach ground water (Rao et al., 1985; Jury et al., 1987).

Pathogens

The extent to which soil can remove pathogens depends on a variety of factors, including the physical, chemical, and biological characteristics of the soil, the size and nature of the organism, and the environmental conditions. The largest organisms, such as protozoa and helminths, are removed effectively by filtration unless the soil contains large pores or continuous voids. Bacteria are also filtered, but in addition may adsorb to soil solid material. Viruses are too small to be filtered by most soil pores and are attenuated only by sorption. The latter mechanism is much more pronounced in unsaturated soil than under saturated conditions, for reasons that are not clear at the present time. Speculation about explanations for the increase in sorption under unsaturated conditions have centered on the role of the air-water interface, either because it forces viral particles nearer to the solid surfaces (Lance and Gerba, 1984) or because the interface itself can trap the particle and perhaps deform it enough to cause inactivation (Powelson et al., 1990).

The most important factor in microorganism survival in soil and ground water is temperature. Below 4°C, microorganisms can survive for long periods of time, but they die off rapidly with increasing temperature. The bacterial die-off rate approximately doubles with each 10°C increase in temperature, but viruses have been observed to follow a more linear relation between inactivation rate and temperature (Yates and Yates, 1987). The soil pH affects survival, which shortens under acidic conditions. There is evidence that viral sorption increases at low pH, because the isoelectric points of soil minerals generally are lower than those of viruses, so that the viral particles become more electropositive than the soil at low pH. Soil moisture affects survival of all microorganisms, with inactivation decreasing significantly as water content is increased above air dry levels. Viral movement is not sufficiently well understood to model at the present time, in part because independent measurements of sorption in the laboratory seriously underestimate the extent of attenuation during transport through unsaturated soil. Gerba and Goyal (1985) discuss the major factors affecting viral movement in soil.

Evidence from monitoring at SAT sites suggests that the larger pathogens are mostly removed by filtration and do not reach ground water during their lifetime in soil, unless a continuous coarse-textured or void pathway is present through the surface soil zone. The greatest removal occurs in the surface mat of suspended particles that forms on the top few millimeters of soil and strains the entering solution.

Field measurements of viral movement at ground water recharge sites are limited. At the Flushing Meadows surface spreading site in Phoenix, Arizona, viruses were not detected after 10 to 30 m (33 to 98 ft) of travel through the fine sandy soil of the system, while more sensitive measurements at the Sweetwater surface spreading site in Tucson, Arizona, detected migration of viruses through the soil system (Powelson and Gerba, 1993). A review of 6 land application (irrigation) sites and 11 rapid infiltration sites by Gerba and Goyal (1985) indicated that viruses have been isolated underground after various migration distances. Vertical migration ranged from 1.4 to 30.5 m (4.6 to 100 ft) at spray irrigation sites and from 2.4 to 67 m (8 to 220 ft) at rapid infiltration sites. At the rapid infiltration sites reviewed by Gerba and Goyal (1985), horizontal migration of viruses in the underground ranged from 3 to over 400 m (10 to over 1,300 ft). Disinfection of the water prior to recharge by surface spreading or injection can minimize or eliminate microorganisms underground, as can proper site selection and management of surface spreading systems.

Disinfection By-products

In most cases, source water intended for recharge will be disinfected with chlorine, chloramine, ozone, or a combination of these disinfectants. Table 3.5 lists the major potentially problematic by-product compounds that would be most likely to be produced from these disinfectants (Bull and Kopfler, 1991). Because bromide is present in many ground water systems, many brominated derivatives are included although data on occurrence frequency and concentrations for these compounds is limited. In addition, Table 3.5 presents only examples of DBP categories by structural class; many others could be listed in most cases.

Nature of Reduced Carbon Species

A major uncertainty arises from the fact that most of the product identification work has been done on standardized humic or fulvic acid fractions isolated from surface water sources of sufficient quality (low organic carbon levels) to be considered as a traditional water supply source. The chemical composition of alternative water sources, however, may be drastically different from that of traditional water sources. In the case of drainage from low-lying peat areas, it might be reasonable to assume that this water source differs only in the higher level of humic carbon present. However, Amy et al., (1984) found that higher levels of trihalomethanes (THMs) are characteristic of more than 200 agricultural drains in the Sacramento River delta and that the larger amount of dissolved organic carbon (DOC) in these water sources is characterized by a higher molecular weight and greater THM reactivity than that found in delta tributaries. Some research showed THM formation potential values as high as 3,580 micrograms per liter ($\mu g/l$) and di- and trichloracetic acid values of 1,650 and 1,990 $\mu g/l$, respectively (Amy et al., 1984). Clearly, carbon level is the largest single parameter controlling DBP levels, although chlorine to carbon dose ratio, pH, temperature, and contact time are also important. However, it remains uncertain whether the carbon levels from non-traditional sources are more or less sensitive

TABLE 3.5 Examples of Problematic[a] Disinfection By-products in Water Prior to Recharge

DBP	Disinfectants	Estimated Fate During Recharge
Chloroform	Chlorine, chloramines	Volatilization, sorption
Bromodichloromethane	Chlorine, ozone	Volatilization, sorption
Trichloroacetic acid	Chlorine	Mineral sorption, ion exchange
Dibromoacetonitrile	Chlorine	?
Formaldehyde	Ozone, chlorine	?
1,1,1-Trichloropropanone	Chlorine	?
MX[b]	Chlorine	Sorption, degradation
Bromate	Ozone	Reaction with soil

[a]Listed DBPs are problematic the sense that the compound commonly has been found with the indicated disinfectant and is of toxicological significance.
[b]MX=3-Chloro-4(d-chloromethyl-5-hydroxy-2(5H)-furanone.

to disinfectant dose, and whether unique by-products are produced from some sources and not others.

The nature of the remaining DOC is also affected by the exposure to disinfectant oxidants. Liao (1983) has shown that dissolved carbon made from purified humic preparations is converted to smaller and/or more polar sizes by exposure to permanganate, and to some degree by hydrolysis. Anderson et al. (1985) also showed that remaining humic materials after low dosages of ozone are more polar and of lower molecular weight than untreated humic material. The effect of such changes on soil retention, metal binding capacity, interactions with other organics, and so on, is unknown. However, one might speculate that this partially oxidized carbonaceous material might be less retentive on soil organic matter because of decreased hydrophobicity and more retentive on soil mineral phases because of increased charge density. In addition to the known DBPs, residual halogen is present when chlorine is used. This material may then appear as adsorbable organic halogen (AOX) and be more biodegradable in anaerobic underground environments than nonhalo- genated organic material.

Stability of DBPs in the Soil and Underground Environment

If ultraviolet disinfection is not used or if chlorine is used in conjunction with ultraviolet radiation, the question of the stability of DBPs in ground water systems becomes important. Changes in the DBP chemical composition of recharge water will occur in the infiltration basin and in the soil vadose zone for surface recharge operations, and in the aquifer during transport and storage, regardless of the input procedure. In surface recharge operations, the water to be recharged will undergo some chemical change due to microbiological activity in the infiltration basin. As organisms respond to the DOC and begin growth cycles, total organic carbon (TOC) may increase or decrease slightly, depending on retention time and the acclimation phase (Bouwer, 1984). In the vadose zone, hydrophobic compounds will be retarded by the soil organic material and subjected to microbial decomposition. Halogenated organic compounds, including the nonvolatile DBPs, will have greater hydrophilicity and correspondingly less retention on soil organic matter and more rapid transport to the water table. Some DBPs, such as the halogenated acetic acids, will behave as fully ionized organic anions at the pH values of water in the vadose zone and may be expected to move at the same speed as the subsurface water. Thus, it might be expected that the smaller halogenated materials will persist longer and penetrate more deeply into the soil horizons than the larger and more hydrophobic molecules. The degree of microbial utilization is most difficult to predict and will probably vary from site to site.

The bioremediation literature notes that trichloroacetic acid is an intermediate in the oxidative degradation of vinyl chloride produced from anaerobic degradation of trichloroethylene and that it does not build up in the soil. This suggests that complete mineralization is possible for this DBP, at least in systems with fully adapted microorganisms. The same is probably true for the other haloacetic acids. Thus, control of the redox environment in the vadose zone through optimizing application rates is probably an important management criterion for these DBPs.

Of the list of DBPs in Table 3.5 one might expect significant losses of formaldehyde and chloroform in the infiltration basin due to volatilization, with somewhat less significant losses of any bromodichloromethane and 1,1,1-trichloropropanone, owing to lower volatilities. Losses of trichloroacetic acid may be expected in the vadose zone.

MX presents an unusual case because its polarity depends on whether the molecule is in the open form, which is a substituted butanoic acid, or in the closed form, which is a substituted hydroxyfuranone. A pH value of less than 5 would give the closed form, which is substantially less polar and therefore more interactive with soil organic matter and more likely to be retained for microbial utilization. In the open acid form, it might be expected to behave like trichloroacetic acid. In addition, MX and its congeners are probably reactive with reduced forms of sulfur in the subsurface environment, but their reaction products are probably of less toxicological significance.

Singer et al. (1993) have studied the fate of chlorinated DBPs when treated and disinfected drinking water was injected and stored in an aquifer at five sites. These authors concluded that THMs and haloacetic acids (HAAs) are removed from chlorinated drinking water during aquifer storage. HAA removal apparently precedes THM removal, and the more highly brominated species tend to be eliminated earliest. Removal of DBPs during aquifer storage was accompanied by a decrease in dissolved oxygen, implicating anaerobic biological mechanisms. In addition to DBP removal, reductions in the concentrations of organic precursor material were observed and correlated with measured reductions in THM formation potential and HAA formation potential. No effort was made to establish the identities of THM or HAA degradation products.

The results from Singer et al. (1993) concur with those from Roberts (1985) who evaluated organic contaminant behavior during a recharge project in Palo Alto, California. He also found that THM losses were correlated with losses of dissolved oxygen. In addition, he found that the loss rate for THMs was nearly 10 times as rapid as the rate of concentration decrease for compounds containing two carbon atoms, represented by trichloroethylene, tetrachloroethylene, and 1,1,1-trichloroethane.

For these reasons, disinfection to control viruses or other pathogens prior to infiltration may not necessarily result in an increase of DBPs in the ground water aquifer. However, it is possible to overload the removal mechanisms, and this is driven by TOC levels in the source water primarily. To provide disinfection and prevent DBP buildup in the ground water it would be necessary to pretreat source water in order to reduce levels of TOC. On the other hand, if the pathogen content of the ground water is not of primary concern, then disinfection (and DBP formation) may be avoided prior to recharge.

SUSTAINABILITY OF THE SAT SYSTEM

Sustainability refers to the long-term viability of an SAT operation, and specifically to the ability of the soil and aquifer to receive recharge water indefinitely without suffering deleterious side effects. Within the vadose zone, clogging is an inevitable side effect of SAT procedures, but is largely manageable with periodic drying to oxidize the accumulated organic material and restore the infiltration rates, along with periodic physical removal of the clogging layer by scraping, raking, or other techniques (Bouwer, 1985).

Of the material that is not completely removed from the recharge water before reaching the aquifer, viruses are of special concern because they may cause disease or pose an unacceptably high risk even when present in low concentrations. Sustainability must therefore include considerable travel distance between the receiving basin of the aquifer and any route to drinking water supplies to allow sufficient viral inactivation to occur (Yates et al., 1986).

Nitrates that enter the aquifer should be regarded as nondegrading, and therefore the recharge operation must operate very efficiently with respect to nitrogen

RAPID INFILTRATION-EXTRACTION PROJECT IN COLTON, CALIFORNIA

The cities of San Bernardino and Colton, California, are required to filter and disinfect their secondary effluent prior to discharge to the Santa Ana River, which is a source of drinking water and is used for body contact recreation. The two cities joined under the auspices of the Santa Ana Watershed Project Authority (SAWPA) to seek a regional solution to their wastewater treatment requirements. They hoped to develop a cost-effective alternative to conventional tertiary treatment (chemical coagulation, filtration, and disinfection) that would still result in an effluent that was essentially free of measurable levels of pathogens. SAWPA conducted a one-year demonstration project to examine the feasibility of a rapid infiltration-extraction (RIX) process to treat unchlorinated secondary effluent prior to discharge to the river and determine whether or not the RIX process is equivalent—in terms of treatment reliability and quality of the water produced—to the conventional tertiary treatment processes specified in the California Department of Health Services Wastewater Reclamation Criteria.

The demonstration was conducted on a site in Colton, California, adjacent to the Santa Ana River bed. Infiltration basins were built at two sites on the property to allow testing the RIX system under a variety of operating conditions. The soils were coarse sands with clean water infiltration rates of about 15 m/day (50 ft/day). Forty-four monitoring wells were sampled for a number of organic, inorganic, microbiological, and physical measurements.

The study indicated that the optimal filtration rate was 2 m/day (6.6 ft/day) with a wet to dry ratio of 1:1 (1 day of flooding to 1 day of drying). Mounding beneath the infiltration basins ranged from 0.6 to 0.9 m (2 to 3 ft), and infiltrated wastewater migrated up to 24.4 m (80 ft) vertically and over 30.5 m (100 ft) laterally before it was extracted and had an aquifer residence time of 20 to 45 days, depending on the recharge site. Extraction of 110 percent of the volume of infiltrated effluent by downgradient extraction wells effectively contained the effluent on the RIX site and minimized mixing with regional ground water outside the project area.

The soil-aquifer treatment reduced the concentration of total coliform organisms 99 to 99.9 percent in samples collected approximately 7.6 m (25 ft) below the infiltration basins, and, generally, water from the extraction wells prior to disinfection contained less than 2.2 total coliforms per 100 ml. While viral levels were as high as 316 viruses per 200 l in the unchlorinated secondary effluent applied to the infiltration basins, only one sample of extracted water prior to disinfection was found to contain detectable levels of viruses. In addition, the turbidity of the extracted water generally averaged less than 0.04 NTU (Foreman et al., 1993).

Although the RIX process greatly reduced microorganism levels in the wastewater, disinfection of the extracted wastewater proved to still be necessary prior to its discharge to the Santa Ana River. Ultraviolet radiation was evaluated as an alternative to chlorination for disinfection. Due, in part, to the high quality of the extracted water, ultraviolet radiation was shown to be effective for the destruction of both bacteria and viruses and will likely be used instead of chlorine in the full-scale project (CH2M Hill, 1992).

removal in the vadose zone where it is transformed. Because proper management of an SAT system has been able to achieve as high as 80 percent removal of nitrogen from the recharge water, control of this chemical may be manageable in long-term operations.

Trace element accumulation in the vadose zone during operation may occur so slowly that exhaustion of the assimilative capacity of the surface zone may not occur in a time that will limit the economic viability of the operation (Chang and Page, 1985); however, the capacity of the soil to remove metal cations is not infinite, and their buildup should be monitored. Site closure at the termination of operation of an SAT system should be regulated to ensure that the metals residing in the surface soil are not disturbed by operations that might change the chemical characteristics of the soil solution and mobilize the contaminants.

Pesticides and other organic chemicals present in recharge water vary significantly in their mobility, persistence, and suspected health effects. Water that contains pesticides known to be resistant to microbial or chemical degradation in soil may pose problems for SAT systems in the long run, because refractory organics are only slowed by solid-phase sorption and not permanently removed from solution.

Ground water recharge operations, either from surface spreading or injection wells, introduce microorganisms to the subsurface environment in addition to organic and inorganic chemicals. Microbial activity associated with artificial recharge may have three principal effects: (1) bacteria may grow near the recharge facility and cause clogging, with a gradual reduction of soil and aquifer hydraulic conductivity; (2) microbial activity may produce substances that adversely affect the taste and odor of recovered water; and (3) pathogenic organisms in the recharge water may travel through the aquifer and cause illness when the water is later recovered for use without disinfection (Ehrlich et al., 1979a). During tests with injection of highly treated sewage into a sand aquifer on Long Island, chlorination of the injectant to 2.5 mg/l suppressed bacterial growth and clogging (Ehrlich et al., 1979b). Similar tests at another injection site showed that clogging could be controlled with a chlorine residual maintained between 0.2 and 1.5 mg/l at the treatment plant (Schneider et al., 1987). The tests of Ehrlich et al. (1979b) also showed that movement of bacteria from the injection well into the aquifer was not extensive. Pathogenic bacteria are generally incapable of competing with soil bacteria for nutrients, while viruses are incapable of reproducing outside of a living cell. They do not multiply and eventually they die and decompose (Bouwer, 1978).

PERFORMANCE AND COMPLIANCE MONITORING

Vadose Zone Monitoring

Ideally, monitoring should be instituted near the point of application in the soil to determine whether the SAT system is functioning in the desired manner. Vadose zone monitoring is difficult, however, for a number of reasons. First, there are essentially only two monitoring methods (soil sampling and vacuum extraction of solution), both of which have drawbacks. Soil sampling is destructive and very localized, so many samples must be taken to give an accurate picture of the average and extreme behavior of the SAT system. This causes substantial land disturbance. Vacuum extraction through porous soil solution samplers permanently installed in the soil is less destructive than soil sampling, but also can produce preferential flow pathways along the sides of the sampling tube if special care is not taken to produce and maintain a tight contact with the surrounding soil. Vacuum extractors also sample an unknown volume of soil and may be very inefficient at intercepting rapidly flowing solution such as might be present in preferential flow channels (Roth et al., 1991). They may also "filter out" microorganisms and chemicals. For these reasons, the contaminant transport characteristics revealed by solution samplers, particularly those located near the surface, might not be representative of the true behavior in the field (Roth et al., 1991).

A second limitation to monitoring near the surface is that the transformation processes occurring within this zone may not have reached completion at the point of monitoring, so interpretation of the information recorded is problematic. Spatial variations in water velocity, chemical concentration, and soil structure are most pronounced near the surface, so the monitoring density required for accurate assessment of average and extreme behavior is greatest in this zone. For these reasons, the best zone of monitoring is generally farther from the surface, such as in the top of the saturated zone. Vadose zone monitoring is generally confined to periodic sampling of the soil and solution to determine general trends.

Tracing of Recharge Water

To follow the migration and fate of recharge water, water quality changes may be monitored by using water samples from wells and lysimeters. Chemical analyses usually focus on indicator chemicals rather than attempting to identify all recharge water chemical species. If a sufficient contrast exists between recharge and ambient waters, then TDS or conductivity may serve as indicators of general migration. Other mobile chemical species that may be used as indicators of recharge waters include chlorides, sulfates, and nitrates. Migration of organic species may be followed by grouping chemicals based on their potential mobility through R values. Lumped parameters such as TOC may serve as a surrogate for wastewater, with values less than 1 mg/l TOC at the wellhead showing adequate dilution and TOC losses for potable use.

Tracing of recharge waters is more difficult if the recharge water quality is similar to that of native formation waters. At the Fred Hervey water reclamation plant in El Paso, Texas, the recharge and formation water are very similar, and a combination of chlorine, nitrogen, and ^{18}O concentrations are used to determine breakthrough of recharge waters at monitoring and recovery wells.

Mixing of Recharge and Ambient Ground Water

The extent of mixing between recharge and ambient ground waters is important in determining whether potentially toxic chemicals in the recharge source waters exceed concentrations that may adversely affect human health when extracted waters are used for potable purposes. The degree to which recharge water has mixed with ambient ground water at a site downgradient from the recharge facility may often be estimated from water quality parameters. If chemical constituents behave conservatively during subsurface migration and mixing, then a linear weighting of concentrations may be used to estimate the degree of mixing (Reeder et al., 1966). For example, chloride and sodium concentrations were used to determine mixing of reclaimed water recharged at East Meadow, Long Island, with the ambient ground water (Schneider et al., 1987). The median sodium and chloride concentrations in the recharge water were approximately 115 and 157 mg/l, respectively, whereas those in ambient ground water were 22 and 27 mg/l, respectively. If the chloride concentration in the ambient ground water is approximately 30 mg/l, in the recharged water is 160 mg/l, and in the observation well sample is C_o mg/l, then the fraction F of recharge water in samples from an observation well can be calculated from the following equation:

$$F = \frac{C_o - 30}{160 - 30} \qquad (8)$$

An observation well sample concentration of $C_o = 100$ mg/l would correspond to a recharge water fraction of 0.54, or 54 percent in the water sampled.

California is developing regulations that place restrictions on the percentage of recharged and reclaimed water to be included in water produced by supply wells for potable use. A supply well can produce water from strata above and/or below as well as from the zone in which subsurface migration of the recharge water is occurring. Also, as a result of radial flow toward the supply well during pumping episodes, part of the water produced by the well may be from parts of the aquifer containing ambient ground water. Under either of these conditions, the water produced by the supply well may be a mixture of recharge water and ambient ground water. Equation (8) can be used to estimate the recharge water fraction in the produced supply water.

Not all chemical constituents allow a simple linear weighting of concentrations to be used to estimate the degree of mixing, especially reactive chemicals and those contributing to the buffering capacity of the system. As an example, calcium may not serve as a good indicator of mixing because of its affinity for clay minerals and because of its role in the calcium-carbon dioxide system. If ambient and recharge waters are mixed within the formation, then calcium carbonate may precipitate, remain in equilibrium with the resulting water, or dissolve, depending on the ratios of carbonate and bicarbonate in the two source waters (Huisman and Olsthoorn, 1983).

Posttreatment of Recharge Waters

For many recharge systems the recharge water tends to lose its chemical identity while in the subsurface, and recovered waters have traditionally required little treatment before use. Even where potable reuse is contemplated, if pretreatment of recharge waters is sufficient, little posttreatment is required. For example, recharge waters in El Paso, Texas, are planned for potable reuse. The moderately weak domestic sewage receives advanced wastewater treatment before injection, and the recovered ground water receives only disinfection before use.

Where treatment is required, posttreatment of recharge waters is no different from ground water treatment. Tertiary treatment can remove taste and odor problems. For removal of organics, aeration (air stripping) and sorption using activated carbon are particularly effective, and neither is inordinately expensive nor difficult to install in a water treatment system. Activated carbon may not remove DOC to sufficiently low values. In that case, membrane filtration or reverse osmosis can be used. Point-of-use water treatment systems may include chlorination, ozonation, or ultraviolet disinfection, ion exchange for water softening, and filtration with activated carbon or membrane filtration for control of organics as well as taste and odor.

SUMMARY

The soil and underlying aquifer have a great capacity to remove chemical contaminants and pathogens from recharge water. The assumption that passage through the soil to the aquifer and through the aquifer to the point of withdrawal provides no treatment and that all treatment must be provided before recharge or after extraction is overly conservative when applied to most chemicals and microorganisms. Ground water recharge and recovery systems can provide significant treatment benefits at relatively low costs and in appropriate circumstances can make use of water of impaired quality attractive.

The ideal soil for an SAT system balances the need for a high recharge rate, which occurs in coarse-textured soils, with the need for efficient contaminant adsorption and removal, which are better in fine textured soils. Because structured soils are undesirable for obvious reasons, the best choice of soil texture for SAT is a structureless fine sand or sandy loam.

The vadose zone has the capacity to remove many constituents of concern during passage of recharge water toward the underlying aquifer. Nitrogen, for example, quickly transforms to nitrate, which is very mobile under normal conditions in the soil but can be removed only by denitrification under anaerobic conditions. This reaction can be enhanced by the proper combination of alternate wetting and drying cycles. Phosphorus levels are reduced by sorption and precipitation, but not completely. Trace metals, with the exception of boron and arsenic, are strongly attenuated and precipitated in the soil. There is some concern about their eventual passage through a soil that has been under SAT for many years or after closure of a site.

Organic chemicals are removed to varying extents by volatilization or chemical or biological degradation during passage through the vadose zone. Some pathogen removal in soil occurs by filtration in the surface clogging mat for the largest organisms and by sorption for bacteria and viruses. Viruses are considerably more mobile in soil than the larger pathogens, although they inactivate in soil eventually. Traditional disinfection by chlorination produces DBPs, which are mobile in soil and persistent to varying degrees.

With proper management and pretreatment, an SAT operation employing surface wastewater spreading with periodic drying to reduce clogging should be sustainable indefinitely. Slow trace element migration remains a concern, however, that requires monitoring during SAT use and regulation after closure. Although near-surface monitoring is desirable for proper vigilance, soil variability makes it difficult to achieve complete coverage with existing devices. Therefore, a combination of periodic near-surface and distant monitoring is important. With adequate management and monitoring, SAT systems can reduce pretreatment costs.

REFERENCES

American Society of Civil Engineers. 1961. Recharge and withdrawal. Committee on Ground Water. ASCE Man. Eng. Pract. 40:72-92.

Amy, G. L., P. A. Chadki, and P. H. King. 1984. Chlorine Utilization during formation of THM in presence of ammonia and bromide. Environ. Sci. Technol. 18:781-786.

Anderson, L. J., J. D. Johnson, and R. F. Christman. 1985. Reaction of ozone with isolated aquatic fulvic acid. Organic Geochemistry. 8:65-69.

Armstrong, D. E., and J. Konrad. 1974. Nonbiological degradation of pesticides. In Pesticides in Soil and Water, W. Guenzi, ed. Madison, Wisc: Soil Science Society of America.

Asano, T. 1985. Artificial Recharge of Groundwater. Boston, Mass: Butterworth.

Bear, J. 1972. Dynamics of Fluids in Porous Media. New York: Elsevier.

Beven, K., and P. Germann. 1982. Macropores and water flow in soils. Water Res. Res. 18:1311-1325.

Bouwer, E. J., P. L. McCarty, H. Bouwer, and R. C. Rice. 1984. Organic contaminant behavior during rapid infiltration on secondary wastewater at the Phoenix 23rd Avenue Project. Water Res. 18(4):463-472.

Bouwer, H. 1978. Groundwater Hydrology. New York: McGraw-Hill.

Bouwer, H. 1984. Wastewater renovation in rapid infiltration systems. Ground Water 22:696-705.

Bouwer, H. 1985. Renovation of wastewater with rapid-infiltration land treatment systems. Pp. 249-282 in Artificial Recharge of Groundwater, T. Asano, ed. Boston, Mass.: Butterworth.

Bouwer, H. 1991. Simple derivation of the retardation equation and application to preferential flow and macrodispersion. Ground Water 29(1):41-46.

Bull R. J., and F. C. Kopfler. 1991. Health effects of Disinfectants and Disinfection By-products. Denver: Am. Water Works Assoc. Res. Found.

Butters, G. L., and W. A. Jury. 1989. Field scale transport of bromide in an unsaturated soil. II. Dispersion modeling. Water Res. Res. 25:1582-1589.

Chang, A. C., and A. L. Page. 1985. Soil deposition of trace metals during groundwater recharge using surface spreading. Pp. 609-626 in Artificial Recharge of Groundwater, T. Asano, ed. Boston, Mass.: Butterworth.

CH2M Hill. 1992. UV Disinfection Pilot Study: Rapid Infiltration/Extraction (RIX) Demonstration Study. Prepared for the Santa Ana Watershed Project Authority, City of San Bernardino, and City of Colton. Santa Ana, Calif.

Crites, R. W. 1985. Micropollutant removal in rapid infiltration. Pp. 597-608 in Artificial Recharge of Groundwater, T. Asano, ed. Boston, Mass: Butterworth.

Dagan, G. 1986. Statistical theory of groundwater flow and transport: Pore to laboratory, laboratory to formation, and formation to regional scale. Water Resour. Res. 22:120S-134S.

Ehrlich, G. G., E. M. Godsy, C. A. Pascale, and J. Vecchioli. 1979a. Chemical changes in an industrial waste liquid during post-injection movement in a limestone aquifer, Pensacola, Florida. Ground Water 17(6):562-573.

Ehrlich, G. G., H. Ku, J. Vecchioli, and T. Ehlke. 1979b. Microbiological effects of recharging the Magothy Aquifer, Bay Park, New York, with tertiary-treated sewage. U.S. Geol. Surv. Prof. Paper 751-E.

Foreman, T. L., G. Nuss, J. Bloomquist, and G. Magnuson. 1993. Results of a 1-year rapid infiltration/extraction (RIX) demonstration project for tertiary filtration. Pp. 21-36 in proceedings of the Water Environment Federation 66th Annual Conference and Exposition, Oct. 3-7, 1993, Anaheim, Calif. Alexandria, Va.: Water Environment Federation.

Gerba, C., and S. Goyal. 1985. Pathogen removal from wastewater during groundwater recharge. Pp. 283-318 in Artificial Recharge of Groundwater, T. Asano, ed. Boston, Mass.: Butterworth.

Ghodrati, M., and W. A. Jury. 1990. A field study using dyes to characterize preferential flow of water. Soil Sci. Soc. Am. J. 54:1558-1563.

Hamaker, J. W., and J. M. Thompson. 1972. Adsorption. Pp. 49-144 in Organic Chemicals in the Soil Environment, C. A. I. Goring and J. W. Hamaker eds. New York: Marcel Dekker.

Hillel, D. 1987. Unstable flow in layered soils; a review. Hydrological Processes 1(2):143-147.

Howard, P. H. 1990. Handbook of Environmental Fate and Exposure Data for Organic Chemicals Vol I and II. Chelsea, Mich: Lewis.

Howard, P. H., R. Boethling, W. Jarvis, W. Meylan, and E. Michalenko. 1991. Handbook of Environmental Degradation Rates. Chelsea, Mich: Lewis.

Huisman, L., and T. N. Olsthoorn. 1983. Artificial groundwater recharge. Boston, Mass.: Pitman.

Jenne, E. A. 1968. Controls on Mn, Fe, Co, Ni, Cu, and Zn concentrations in soils and water. Pp. 337-387 in Trace Organics in Water, R. F. Gould ed. Advances in Chemistry Series, Vol. 73. Washington, D.C.: American Chemical Society

Jury, W. A. 1985. Spatial Variability of Soil Physical Parameters in Solute Migration: A Critical Literature Review. EPRI Topical Rep. E4228, Electric Power Research Institute. Palo Alto, Calif.

Jury, W. A., and H. Fluhler. 1992. Transport of chemicals through soil: Mechanisms, models, and field applications. Adv. Agron. 27:141-201.

Jury, W. A., and K. Roth. 1990. Evaluating the role of preferential flow on solute transport through unsaturated field soils. Think Tank Workshop Report. Pp. 23-30 In Field Scale Water and Solute Flux in Soils, K. Roth et al., eds. Basel, Switzerland: Birkhaeuser.

Jury, W. A., W. F. Spencer, and W. J. Farmer. 1983. Model for assessing behavior of pesticides and other trace organics using benchmark properties. I. Description of model. J. Environ. Qual. 12:558-564.

Jury, W. A., W. J. Farmer, and W. F. Spencer. 1984a. Model for assessing behavior of pesticides and other trace organics using benchmark properties. II. Chemical classification and parameter sensitivity. J. Environ. Qual. 13:567-572.

Jury, W. A., W. F. Spencer, and W. J. Farmer. 1984b. Model for assessing behavior of pesticides and other trace organics using benchmark properties. III. Application of screening model. J. Environ. Qual. 13:573-579.

Jury, W. A., W. F. Spencer, and W. J. Farmer. 1984c. Model for assessing behavior of pesticides and other trace organics using benchmark properties. IV. Review of experimental evidence. J. Environ. Qual. 13:580-585.

Jury, W. A., D. D. Focht, and W. J. Farmer. 1987. Evaluation of pesticide groundwater pollution potential from standard indices of soil-chemical adsorption and biodegradation. J. Environ. Qual. 16:422-428.

Jury, W. A., D. Russo, G. Streile, and H. Elabd. 1990. Evaluation of volatilization by organic chemicals residing below the soil surface. Water Resour. Res. 26:13-20.

Jury, W. A., W. R. Gardner, and W. H. Gardner. 1991. Soil Physics. New York: JohnWiley. 239 pp.

Karlson, U., and W. T. Frankenberger. 1989. Removal of selenium from evaporation pond sediments by biological transformation. Sci. Total Environ. 92:41-54.

Kung, K.-J. S. 1990. Preferential flow in a sandy vadose zone: 1. Field observation. 2. Mechanism and implications. Geoderma 46:51-72.

Lance, J. C., and C. P. Gerba. 1984. Virus movement in soil during saturated and unsaturated flow. Applied Environmental Microbiology. 47:335-337.

Liao, W. T., R. F. Christman, J. D. Millington, and J. R. Hass. 1982. Structural characterization of aquatic humic material. Environ. Sci. Technol. 16:403-410.

Liss, P. S., and P. G. Slater. 1974. Fluxes of gases across the air-sea interface. Nature 247:181-184.

Nellor, M. A. 1980. Health effects of water reuse by ground water recharge. Tech. Rep. CRWR-175. Center for Research in Water Resources, University of Texas, Austin.

Omoti, U., and A. Wild. 1979. Use of fluorescent dyes to mark the pathways of solute movement through soils under leaching conditions. 2. Field experiments. Soil Sci. 128:93-104.

Powelson, D. K., and C. P. Gerba. 1993. Virus transport and removal in wastewater during aquifer recharge. Water Research. 27:583-590.

Powelson, D. K., J. R. Simpson, and C. P. Gerba. 1990. Virus transport and survival in saturated and unsaturated flow through soil columns. J. Environ. Qual. 19:396-401.

Pratt, P. F., W. W. Jones, and V. E. Hunsaker. 1972. Nitrate in deep soil profiles in relation to fertilizer rates and leaching volume. J. Environ. Qual. 1:97-102.

Rao, P. S. C., A. G. Hornsby, and R. E. Jessup. 1985. Indices for ranking the potential for pesticide contamination of groundwater. Proc. Soil Crop Soc. Fla. 44:1-8.

Reeder, H., W. Wood, G. Ehrlich, and R. Sun. 1966. Artificial recharge through a well in fissured carbonate rock, West St. Paul, Minnesota. U.S. Geological Survey Water Supply Paper 2004.

Ritchie, J. T., D. E. Kissel, and E. Burnett. 1972. Water movement in undisturbed swelling clay soil. Soil Sci. Soc. Am. Proc. 36:874-879.

Roberts, P. V. 1985. Field observations of organic contaminant behavior in the Palo Alto baylands. Pp. 647-679 in Artificial Recharge of Groundwater, T. Asano, ed. Boston, Mass: Butterworth.

Roth, K., W. A. Jury, H. Fluhler, and W. Attinger. 1991. Transport of chloride through an unsaturated field soil. Water Resour. Res. 27:2533-2541.

Schneider, B. J., H. Ku, and E. Oaksford. 1987. Hydrologic effects of artificial recharge experiments with reclaimed water at East Meadow, Long Island, NY. U.S. Geol. Surv. Water Resour. Inv. Rep. 85-4323.

Scotter, D. R., and P. Kanchanasut. 1981. Anion movement in a soil under pasture. Austr. J. Soil Res. 19:299-307.

Sharpless, R., E. Wallihan, and F. Peterson. 1969. Retention of zinc by some arid zone soils treated by zinc sulfate. Soil Sci. Soc. Am. Proc. 33:901-904.

Singer, P. C., R. D. G. Pyne, M. AVS, C. T. Miller, and C. Mojonnier. 1993. Examining the impacts of aquifer storage and recovery on DBPs. J. American Water Works Assn. 85(11):85-94.

Sposito, G. 1981. The Thermodynamics of Soil Solutions. Oxford: Oxford University Press.

Thomas, R. G. 1982. Volatilization from water. Chap. 15 in Handbook of Chemical Property Estimation Methods, W. J. Lyman et al., eds. New York: McGraw-Hill.

Vecchioli, J., H. F. H. Ku, and D. J. Sulam. 1980. Hydraulic Effects of Recharging the Magothy Aquifer, Bay Park, New York, With Tertiary-Treated Sewage. Geol. Surv. Prof. Paper 751-F. U.S. Department of the Interior, U.S. GPO, Washington, D.C.

Yates, M. V., and S. R. Yates. 1987. Modeling virus survival and transport in the subsurface. J. Contam. Hydrol. 1:329-345.

Yates, M. V., S. R. Yates, A. Warrick, and C. Gerba. 1986. Predicting virus fate to determine septic tank setback distances using geostatistics. Appl. Environ. Microbiol. 52:479-483.

4

Public Health Issues

A major consideration in the use of impaired quality water for artificial recharge is the possible presence of chemical and microbiological agents in the source water that may be hazardous to human health. Such concerns apply primarily to potable use, although human exposure may occur from nonpotable uses such as agricultural irrigation. In general, however, the potential for exposure to possible hazards is less for nonpotable reuse, and consequently the risks are significantly lower. Thus this chapter focuses primarily on health issues related to potable reuse.

The traditional maxim for selecting drinking water supplies has been to use the highest quality source available. This principle has guided the selection of potable water supplies for at least 150 years. Thus, although indirect potable reuse occurs throughout the nation and world wherever treated wastewater is discharged into a water course or underground and withdrawn downstream or downgradient for potable purposes, such sources are in general less desirable than using a higher quality drinking water source. The central question for artificial recharge then is: Can ground water recharged with source waters of impaired quality satisfy this maxim? If so, can users be assured that such waters do not threaten human health? How much and what kind of treatment is required prior to recharge? Was treatment planned considering (1) regulatory requirements which may prohibit ground water contamination, (2) the extent to which improvement in quality can occur in the soil and aquifer because of chemical and microbiological transformations, (3) whether the extracted water can meet drinking water standards as set by EPA, and (4) whether the extracted water will be used for potable purposes or other purposes that might result in human exposure?

Public health concern over the use of recovered water from ground water recharged with source waters of impaired quality centers on the difficulty in identifying and estimating human exposures to the potentially toxic chemicals and microorganisms that may be present. To some extent the assessment of possible health risks can rely on the vast body of knowledge that has been developed for water supplies using conventional source waters, such as ground water from relatively uncontaminated aquifers and surface waters. However, there is a substantial amount of uncertainty even for such waters, principally related to the presence of synthetic organic chemicals, inorganic chemicals disinfection by-products, and pathogenic organisms.

Studies have been made of the chemical and microbiological characteristics of recovered water, although they are limited in number and scope. Several studies have shown that the recovered water can meet drinking water standards, even when the recharge source is treated municipal wastewater. Such findings lead some experts to the conclusion that these extracted waters are as acceptable as water supplied from traditional sources. Other experts strongly disagree, saying that water originating from an impaired source is inherently more risky. For instance, disinfection of the recharge waters may develop a different mix of disinfection by-products (DBPs), often unidentified, from those found in conventional water supplies. Also, the characterizations of the organic material and the full range of microbiological constituents are incomplete. In addition, source waters of impaired quality and recharge water withdrawn from the aquifer at the point of use may contain some contaminants at higher concentrations than are likely to be present in conventional water supplies. And throughout the whole process there is increased reliance on technology and management, leaving open the door for errors. Thus, the question arises whether drinking water standards developed for conventional water supply systems are sufficiently protective of human health when ground water is recharged with waters of impaired quality.

The assessment of health risks associated with recharge using impaired sources is far from definitive because there are limited chemical and toxicological data and inherent limitations in the available toxicological and epidemiological methods. The limited data and extrapolation methodologies used in toxicological assessments provide a source of limitations and uncertainties in the overall risk characterization. Similarly, epidemiological studies suffer from the need for very long time periods required, because cancers have latency periods of 15 years or more. Also, such studies require large populations to uncover the generally low risks associated with low concentrations of toxicants. Past studies of the possible adverse health effects from reclaimed water have tended to be limited in terms of toxicological characterization and have focused on those chemicals for which drinking water standards exist.

The challenge in considering the health risks from recharge systems is to assess and understand these relative risks and develop strategies for the use and operation of recharge systems to minimize them. A primary goal is the need to minimize the concentrations of possible DBPs and the potential exposures to pathogenic microorganisms. This goal takes on added dimensions, however, in recharged ground water systems because many of the organic precursors of the DBPs are different generally from those in conventional water supply systems. In addition, the behavior and fate of the microorganisms, as well as the DBPs (and, indeed, other chemical toxicants) in the ground water recharge system affect their concentrations at the point of extraction. Developing an understanding of the chemical and microbiological composition and changes in these complex recharge systems will allow the optimal use of this water augmentation strategy.

RISK ASSESSMENT METHODOLOGY, APPROACHES, AND INTERPRETATION

Human health risk assessment is a process used to evaluate the nature and magnitude of potential health risks associated with exposure to environmental agents, including chemicals and microorganisms. The product of the evaluation is a statement regarding the probability that populations so exposed will be harmed, and to what degree—whether expressed in quantitative or qualitative ways (NRC, 1994). Such assessments generally contain four steps (NRC, 1983): hazard identification, dose-response assessment, exposure assessment, and risk characterization. This methodology has been used extensively to characterize the risks associated with environmental and occupational hazards and can be used along with epidemiological information to provide a perspective on the possible risks related to exposures from the use of ground water recharged with waters of impaired quality.

For recovered water, hazard identification entails identification of the contaminants that are suspected to pose health hazards and a description of the specific forms of toxicity (neurotoxicity, carcinogenicity, and so on) that can be caused by the contaminants of concern. Information for this step is typically derived from epidemiological and animal studies and other types of experimental work. Carcinogenic properties and noncarcinogenic effects are considered.

Dose-response assessment entails a further evaluation of the conditions under which the toxic properties of a chemical might be manifested in exposed people, with particular emphasis on the quantitative relation between the dose and the toxic response. The development of this relationship may involve the use of mathematical models. A dose-response assessment identifies any toxicological endpoints associated with specified exposure levels or provides an estimate of the relationship between the increased likelihood and/or severity of adverse effects and the extent of exposure to a chemical. (A discussion of the issues pertinent to a microbial risk assessment can be found later in this chapter.)

Exposure assessment involves specifying the population that might be exposed to the agent of concern, identifying the routes through which exposure can occur, and estimating the magnitude, duration, and timing of the doses that people might receive as a result of their exposure.

Risk characterization involves integration of information from the first three steps to develop a qualitative or quantitative estimate of the likelihood that any of the hazards associated with the agent of concern will be realized in exposed people. This is the step in which risk assessment results are expressed. Risk characterization should also include a full discussion of the uncertainties associated with the estimates of risk.

Not every risk assessment encompasses all four steps. Sometimes only a hazard identification will be conducted to evaluate the potential of a substance to cause human health effects. Regulators sometimes take the additional step of ranking the potency of chemicals—what is known as hazard ranking. Sometimes potency information is combined with exposure data to produce a risk ranking. These techniques all use some, but not all, of the four steps of the quantitative risk assessment process (NRC, 1994).

For the purpose of assessing chemical constituents in recovered water, no original toxicity evaluation is done. Instead, reference toxicity values such as reference doses (for noncarcinogens) and potency factors (for carcinogens) are used. The reference dose for a noncarcinogenic end point is based on the assumption that a threshold exists for that specific toxic effect. It is an estimated dose for a daily exposure that is likely to be without an appreciable risk of deleterious effects during a lifetime. For carcinogens, cancer potency (or slope) factors are usually estimated using a linear nonthreshold mathematical extrapolation model for low-dose extrapolation. The potency factor is characterized as an upper-bound estimate or the 95th percentile confidence limit (95% UCL) of the probability of response per unit intake of chemical over a lifetime. The upper-bound estimate means that the actual risk is likely to be less than the predicted risk.

The potential doses to the exposed population are calculated on the basis of known or estimated parameters. For recovered water to be used for potable purposes, the exposure is assumed to be the dose contained in 2 liters (l) of water that is consumed by an adult on a daily basis for a lifetime. However, it is recognized that there may be additional routes of exposure, such as skin absorption and inhalation of volatile chemicals from water used indoors.

The potential health risks associated with the source water are characterized according to the nature and magnitude of the risk. For carcinogens, the risk is expressed as the probability of cancer occurrence. The negligible risk level generally recognized by regulatory agencies such as the Environmental Protection Agency (EPA) is 1 in 1 million (1×10^{-6}). This means that there is the probability that one person in a population of 1 million will get cancer at the estimated exposure level. This also means that, for example, at the present background rate of 25 percent cancer occurrence in the U.S. population, exposure to the chemical will increase the risk of getting cancer from 250,000 in 1 million to 250,001 in 1 million.

For noncarcinogens, if the exposure (or dose) is less than the reference dose, then it is not likely to be associated with health risk. As an indirect means to characterize exposure, one may compare the concentration of a chemical found in the source water to the drinking water standard established for that chemical. The comparison will be valid if the standard is derived from a reference dose and the exposure to an adult is equivalent to a dose resulting from consuming 2 liters of the water per day.

Another approach is to use the hazard index for assessing the overall potential for noncarcinogenic effects posed by chemicals in the source water. The underlying assumption is that multiple subthreshold exposures could result in an adverse effect and that the magnitude of the effect will be proportional to the sum of the ratios of the subthreshold exposures to acceptable exposures. If the sum of the ratios (or hazard indices) exceeds one, the exposures may result in a potential health effect.

Drinking water standards and health advisories (HAs) are published by the Office of Water of the Environmental Protection Agency (U. S. Environmental Protection Agency, 1993b). The drinking water standards include maximum contaminant level goals (MCLGs) and maximum contaminant levels (MCLs). The MCLGs (which are non-enforceable) are set at a concentration of zero for carcinogens as a matter of EPA policy. However, none of the MCLs (which are enforceable) are zero because they are based on a number of factors, including acceptable risk, detection limits, feasibility, and economic factors. The Health Advisory Program was started by EPA in 1978 to provide information and guidance to individuals and agencies concerned with potential risk from drinking water contaminants for which no national regulations currently exist. HAs are prepared for contaminants that meet two criteria: (1) the contaminant has the potential to cause adverse health effects in exposed humans, and (2) the contaminant is either known to occur or might reasonably be expected to occur in drinking water supplies. Guidance for the first 20 contaminants was issued in 1979, and by 1994 the list contained 189 chemicals and 6 radionuclides.

In assessing the potential impacts of long-term human exposure to recharged ground water, the degree to which any recovered water (after treatment) used as a public water supply meets these (enforceable) MCLs would have to be determined. However, in considering the suitability of a recharge system when the aquifer is to be used as a potable supply, the HAs should be considered as well, especially when there is no comparable MCL for a particular chemical.

STUDIES OF HEALTH IMPACTS

The public health implications of direct and indirect potable use of recovered water have been studied at a number of sites. These include studies of direct potable reuse in Windhoek, Namibia (the only city in the world with a direct potable reuse facility) and Denver, Colorado (where direct reuse was studied extensiely but not adopted); studies of indirect potable reuse via surface sources at San Diego, California, and Tampa, Florida; and studies of indirect reuse via injection in Los Angeles. These studies review varying combinations of chemical characterization of source water, toxicological testing of source water, and epidemiological studies of populations using the recovered water. Most focus on chemical constituents, so limited information is available on impacts from microorganisms.

The major activities that have been conducted to evaluate the health-related aspects of using recovered recharge water were conducted during the period 1975 to 1987 in California, where there is significant potential for reuse. In 1975, the State of California established a Consulting Panel on Health Aspects of Waste Water Reclamation for Groundwater Recharge to recommend a program of research that would assist in the establishment of criteria for ground water recharge to augment public water supplies and help develop programs of reclamation consistent with these criteria. In its report (State of California, 1987), the panel confined its discussions to ground water recharge by surface spreading and concluded that uncertainties exist regarding health effects from the use of reclaimed water primarily due to stable organic materials. To address the uncertainties, the panel recommended that comprehensive studies on health effects of ground water recharge be initiated at existing projects and new demonstration projects. Research was recommended for contaminant characterization, toxicology, and epidemiological studies of exposed populations.

OLAC Water Reuse Study - Montebello Forebay, Los Angeles County, California

In 1978, six water supply and wastewater agencies in Orange and Los Angeles counties organized the OLAC Water Reuse Study. One of the activities of the study was the initiation of a 5-year epidemiological Health Effects Study, guided by the consulting panel's recommendations, which was published (Nellor et al., 1984).

The Health Effects Study was conducted at the Whittier Narrows Ground Water Replenishment Project, located in the Montebello Forebay area of Los Angeles County, where disinfected filtered secondary effluent, stormwater runoff, and imported river water have been used for replenishment since 1962. The study was designed to develop a database that would enable health and regulatory authorities to determine whether the use of reclaimed water for ground water replenishment at the project should be maintained or modified. The research included toxicological studies of ground water, reclaimed water, and other replenishment water supplies to isolate and identify health significant organic constituents, and epidemiological studies of populations ingesting recovered water to determine if their health characteristics differed significantly from those of a demographically similar control population (i.e., a geographical comparison study).

The results of the studies did not demonstrate any measurable adverse impacts on the area's ground water or the health of the population ingesting this water. Specifically, a 1981 household health survey of women residing in the Montebello Forebay study area showed no elevated levels of specific illnesses or other differences in general health. The study was controlled for the potential confounding effects of factors such as smoking, alcohol consumption, bottled water usage, and length of residence. In addition, based on an evaluation of health and vital statistics data for the period from 1969 to 1980, it was reported that residents of the area that received recovered water experienced no increased rates of infectious diseases, congenital malformations, infant and neonatal mortality, low birth weight, cancer incidence, or death due to heart disease, stroke, stomach cancer, rectal cancer, bladder cancer, colon cancer, or all cancers combined, when compared with residents of two control areas that did not receive recovered water.

Concentrated organic residues derived from all replenishment sources and ground waters elicited mutagenic responses in the Ames Salmonella Microsome Mutagen Assey that were related to the presence of a mixture of toxic organic compounds. Chemical and biological assays indicated that low levels of compounds belonging to the two classes of organic halides and epoxides may have contributed to the mutagenicity. The mutagenicity of whole-sample residues was not accounted for. The overall analysis identified approximately 10 percent of the total organic carbon, and the data were not adequate to judge whether or not the majority of compounds present and of greatest health concern were identified. A mammalian cell transformation assay was also used; however, the assay was complicated by problems with fungal contamination and limited number of samples. The mutagenic response of the recovered water samples fell between the surface runoff and imported river water assay values. The limited number of samples from each site and the complexity of the percolation process and aquifer systems precluded a more rigorous statistical analysis of the correlations between the mutagenic response data and the sample sites.

The merits and limitations of the report on the Health Effects Study were analyzed by the Scientific Advisory Panel for Groundwater Recharge and used to prepare the Panel's guidelines (State of California, 1987). Overall, state-of-the-art procedures were used for characterization of the water samples. For inorganics, the water quality data for ground water and recovered source samples were summarized and compared with the existing drinking water standards. Traditional chemical evaluations were performed on water samples in each of four study sites. In general, all sampling well sites showed values within primary drinking water standards; 3 of the 10 Montebello Forebay wells had samples exceeding secondary standards for iron and manganese.

For organic chemicals, the data suggested that a group of nontargeted organics (phthalates, solvents, petroleum by-products) may be more useful markers of future impacts of recovered water than the targeted organics because of their greater concentration in recovered water. Data also suggested that industrial solvents could be used to monitor future impacts of replenishment with reclaimed water. Certain compound groups (phthalates, chlorinated phenols, s-triazine herbicides, phenylacetic acid) may be useful indicators of impacts from replenishment by waters other than reclaimed water. Overall, the data established that ground water in the study sites is currently contaminated with a variety of organic compounds of industrial, and perhaps treatment, origin.

The panel raised the issue of whether Ames mutagenicity assay data are adequate in the absence of other toxicological information to serve as a basis for risk assessment. There was some consensus that the conventional rodent studies would be needed but no clear consensus as to how such studies could or should be carried out with water samples or water sample residues.

Overall, this committee, the National Research Council's Committee on Ground Water Recharge, supports the panel's view that the demographic comparison studies are useful in demonstrating the feasibility of this hypothesis and enabling a rapid assessment of potential threats to public health by using available morbidity, mortality and census information originally collected for other purposes. There is, however, a deficiency because exposure and outcome data describe characteristics of groups, and not individuals. Information is not available on potential confounding factors. The sensitivity of the studies on cancer and chronic disease rates to detect effects of recovered water is severely weakened by the short time period between first exposure to recovered water and the time of the study, especially when compared with the long latency period typically involved (15 years and more) between first exposure and cancer diagnosis and the high in-migration rates in the study area. As for the toxicological data, the present review finds that the Ames data constitute only a very small part of the full toxicological characterization that is needed and, therefore, are not adequate to support a risk assessment of the source water.

Total Resource Recovery Project, City of San Diego, California

The City of San Diego, which imports its water supply from other parts of the state, is projecting a need for additional water for the next decade (Western Consortium for Public Health, 1992). The city is investigating advanced treatment technologies for use of potential new sources—among which is the local municipal wastewater. A 5-year Health Effects Study was conducted for the City of San Diego to investigate whether a proposed wastewater treatment scheme can reliably reduce contaminants of concern to levels such that the health risks to the population are not greater than those associated with the present water supply (Western Consortium for Public Health, 1992). In this study, the risks associated with using the Miramar raw water supply and with the effluent generated from the advanced wastewater treatment (AWT) plant were compared.

When compared with other raw water supplies sampled throughout the United States, concentrations of metals and of the majority of organic constituents detected in AWT and Miramar waters were either in the lower portion of or below the range of averages. The same findings were reported for the local raw wastewater except that values for butylbenzyl phthalate, di-n-octyl phthalate, and bis(2-ethylhexyl) phthalate exceeded the U.S. range. Testing of organic extracts of water in the Ames assay, micronucleus test, 6-thioquanine resistance assay, and cell transformation assay indicated that the AWT water appeared to show less genotoxic or mutagenic activity than the low levels observed in the Miramar water.

An epidemiological study was conducted, which included analysis of the vital statistics of San Diego County women interviewed from 1987 to 1989, and a neural tube defect survey using data from 1978 to 1985 to establish prevalence rates in California and San Diego.

No significant difference for the two areas was found when the annual prevalence-at-birth rates for selected defects (anencephalus and spina bifida combined) were estimated. The study also included characterization of the reproductive health of the women but the results were not reported.

A risk assessment was performed for all compounds whose reference doses (RfD) and/or unit risk values were available in the EPA's Integrated Risk Information System database. These include arsenic, barium, boron, manganese, bromodichloromethane, bromoform, benzoic acid, benzyl butyl phthalate, bis(2-ethylhexyl) phthalate, chloroform, dibromochloromethane, toluene, and trichlorofluoromethane. For carcinogens, the overall mean estimate of lifetime risk from consuming Miramar water at 2 l/day is about 3 in 10,000. Dermal and inhalation absorption resulted in a risk of 1.6 in 10,000. About 98 to 99 percent of this risk is derived from arsenic, and the remaining percentage from trihalomethanes (THMs), which contributed a risk of less than 1 in 100,000. The concentrations of arsenic and THMs in Miramar water are approximately 2 and 10 percent of the drinking water standards, respectively. The mean cancer risk estimate for Miramar water would be 0.9 in 10,000 when the uncertainty is removed by treating the cancer potency value as a point estimate. The cancer risk estimates for the AWT water were 4.8×10^{-7} for dermal, inhalation, and drinking exposures and 8.2×10^{-7} for drinking only.

For noncarcinogens, the hazard index method was used. The chemicals included bis(2-ethylhexyl) phthalate, boron, manganese, and toluene in AWT water and barium, benzyl butyl phthalate, benzoic acid, bromoform, bromodichoromethane, chloroform, dibromochloromethane, manganese, toluene, and trichlorofluoromethane in Miramar water. The hazard indices were 0.077

for AWT and 0.051 for Miramar water. Therefore these chemicals are not anticipated to present a significant health risk.

Tampa Water Resource Recovery Pilot Project, Tampa, Florida

In another study, Florida and the West Coast Regional Water Supply Authority (WCRWSA) investigated the potential indirect reuse of treated AWT effluent generated at the City of Tampa Hookers Point Facility. Toxicological testing was performed on product water produced from a pilot plant (treated municipal wastewater plus filtration) and reference water (treated municipal wastewater). The studies included the following: Ames test, sister chromatid exchange and micronuclei analysis for genotoxicity, 90-day subchronic gavage studies in mice and rats, induction of Strain A lung adenoma, reproductive study in mice and teratological study in rats, and mouse (sencar) skin initiation-promotion study (J. Doull and J. Borzelleca, personal communication, 1992; Pickard et al., undated). The results of these studies were negative. Analytical chemistry results showed that the quality of the pilot plant effluent is as good or better than other sources of raw water such as the reference River water. The pilot plant effluent meets current and proposed EPA drinking water standards and the World Health Organization's guidelines for drinking water quality.

Potable Water Reuse Demonstration Project, Denver, Colorado

The Comprehensive Health Effects Testing Program for the Denver Water Department's Potable Water Reuse Demonstration Project was designed to evaluate the relative health effects of two water types: one was highly treated (including lime precipitation, activated carbon filtration, reverse osmosis, and various filtration and disinfection steps) recovered water derived from secondary treated wastewater and the other was Denver's drinking water (Lauer et al., 1990). The protocol provided for comparative testing of concentrates from Denver's drinking water (from the Foothills Water Treatment Plant), reuse demonstration plant effluent, and reuse demonstration plant effluent treated with ultrafiltration instead of reverse osmosis.

In a 104-week chronic toxicity and carcinogenicity study in Fischer 344 rats and B6C3F1 mice, samples were given at 150 or 500 times the original concentration in drinking water ad libitum. No treatment-related lesions or neoplasms were observed in the animals. In a reproductive/teratology study, rats were administered the test material. No treatment-related effects on reproductive performance, growth, mating capacity, survival of offspring, or fetal development in the animals were found.

Analysis of existing data base

Summary information pertinent to the key health effects studies is displayed in Table 4.1. In each of these studies, efforts were made to ascertain any undesirable toxicological effects of recharging ground water with waters of impaired quality. All of the studies employed state-of-the-art methodologies to measure toxicological effects and to determine the identities of inorganic and organic chemical components. In some studies, additional experimentation was directed at measuring whole-sample mutagenicity and/or the mutagenicity of the fraction of the organic content that was not identifiable. Overall, no significant toxicological properties were found, although methodological limitations in all cases prevent interpretation of this result as indicating that no health effects are associated with human consumption of recovered water from impaired-quality sources.

As is the case with any "not detectable" result in analytical chemistry, when "no observed effect" is found in toxicological and epidemiological studies, that term must be defined. The question is, what is the most sensitive health effect that could have been detected by the methodology employed? Unfortunately, this is a more difficult question for the methods of toxicology and epidemiology than for the methods of analytical chemistry. For example, the Ames Salmonella Microsome Mutagen Assay, in the absence of other toxicological data, is not an adequate basis for the estimation of risk. Conventional animal studies, conducted for a measurable health outcome, would be preferable to complete the spectrum of toxicological testing. Single-chemical and simple in vitro toxicological evaluations are not likely to be responsive to the question of whether the aggregate organic substances in the recovered water would cause any meaningful risk to populations receiving them.

For these reasons, broader evaluation strategies are necessary to estimate health risk. Confidence is gained in the analytical chemistry if one or more target organic molecules of unquestionable health risk are spiked into the sample as external standards, and compounds subject to compliance are specifically examined. Using additional in vitro tests (micronucleus test, 6-thioquanine resistance assay, cell transformation assay) may not be as helpful. Comparing the recharge water with existing or currently acceptable water in terms of genotoxicity evaluation can be helpful. In the San Diego study, for example, AWT effluent showed lower genotoxic or mutagenic activity in the above in vitro tests than the raw water supply. Geographical comparisons can be effective in demonstrating the feasibility of the hypothesis and enabling a rapid assessment of potential threats to public health by using available morbidity, mortality, and census information. The deficiency in these techniques is that exposure and outcome data describe characteristics of groups, not of individuals. A causal relationship is difficult to establish using this method.

To date, the application of state-of-the-art methodology for chemical and limited toxicological analysis has failed to show that professionally managed ground water recharge programs produce extracted water of a lower quality from a health perspective than water from other historically acceptable water sources. But without a complete organic chemical analysis and sufficient animal testing necessary to detect the chemicals and health effects it is impossible to be certain that these waters are suitable or will remain so. In other words, we have reached a point where problems are "not detectable" or "not observable" but we are unable with present data (considering costs) to give any finer meaning to this statement. The Scientific Advisory Panel on Groundwater Recharge (State of California, 1987), as previously discussed, recognized this point and offered several recommendations and criteria for the conduct of future recharge evaluation studies that remain valid today:

- Prospective health surveillance of populations, at least initially, should be part of any project proposing to use reclaimed wastewater to recharge ground water.
- Analysis should emphasize tests of concentrates to determine whether likely harmful substances are present at low levels.
- Single-chemical and simple in vitro toxicological evaluations are not likely to be responsive to the question of whether the aggregate organic substances in recovered water would cause any meaningful risk to populations receiving them. A reasonable assessment can be addressed only by whole animal tests on mixture concentrates and by retrospective surveillance of population. State-of-the-art toxicological studies in animals provide the only recognized methods for evaluating risk prior to human exposure.
- Chemical analysis and monitoring should be continued on reclaimed wastewater as well as extracted ground water to ensure that concentrations of key identified substances, such as those with drinking water standards, are not exceeded and that any other biologically active chemicals are identified.
- The state of the art of chemical and biological monitoring, toxicology, treatment technology, and epidemiology should be reviewed periodically and appropriate adjustments made in project monitoring and operation.

CHEMICAL CONSTITUENTS OF CONCERN

Inorganic Chemicals, Pesticides, and Other Organic Chemicals in Source Waters

Chemicals typically found in irrigation return flow and stormwater runoff were summarized in Table 2.23 in Chapter 2. Table 4.2 presents the concentration ranges of the chemicals identified compared to U.S. drinking water standards (maximum contaminant levels, or MCLs) or other guidance levels (e.g., health advisories), and information on the potential health effects associated with these chemicals. In addition, the concentrations of chemicals found in

TABLE 4.1 Summary of Existing Health Effects Studies

	Types of Water Studied	Chemical Analysis	Health Effects Data
Montebello Forebay, Los Angeles County, California (Nellor, et al., 1984)	Disinfected filtered secondary effluent, storm runoff, and imported river water used for replenishment. Also recovered ground water.	Inorganics and organics. Only 10% of total organic carbons identified. Chemical values were within primary drinking water standards.	Epidemiology: In the geographical comparison study, population ingesting recovered water did not demonstrate any measurable adverse health impact. Household survey (women): No elevated levels of specific illnesses or other measures of general health. Toxicological testing: Concentrated organic residues from all replenishment sources and ground water. Positive mutagenic responses in the Ames assay.
Total Resource Recovery Project, City of San Diego, California (Western Consortium for Public Health, 1992)	Advanced wastewater treatment (AWT) effluent, Miramar raw water supply (current drinking water supply).	Inorganics and organics. Concentrations of metals and majority of organics were either in the lower portion of or below the range of averages of other raw water supplies sampled throughout the United States	Epidemiology: Reproductive health and vital statistics. Results not yet reported. Neural tube defects study: No significant effect. No health risk from chemicals identified based on use of reference doses and cancer potencies.

Tampa Water Resource Recovery Pilot Project, Tampa, Florida (Pickard et al., undated; Doull and Borzelleca, 1992)	Product water (granular activated carbon) from pilot plant. Reference water treated effluent from AWT plant.	Pilot plant effluent met EPA primary drinking water standards and World Health Organization's guidelines for drinking water quality.	Toxicological testing: Ames assay, sister chromatid exchange, micronuclei; 90-day gavage studies in mice and rats; strain A lung adenoma induction; mouse reproductive study; rat teratology study; mouse skin initiation-promotion study. All studies showed negative responses.
Potable Water Reuse Demonstration Project, Denver, Colorado (Lauer et al., 1990)	Concentrates from Denver's drinking water, highly treated recovered water (reuse demonstration plant effluent, and reuse demonstration plant effluent treated with ultrafiltration instead of reverse osmosis).	Inorganic constituents within U.S. drinking water standards. Standards do not exist for some chemicals found.	Toxicological testing: 2-year chronic/carcinogenicity study in rats and mice Rat reproductive/teratology study. No treatment-related effects observed.

Note: See original reports for more detailed information.

TABLE 4.2 Health Effects of Chemicals Identified in Irrigation Return Flow and Urban Stormwater Runoff

Chemical[a]	Health Effects[b]		Carcinogen Classification[c]	Concentration[d] µg/l	Maximum Contaminant Level[e] µg/l (10^{-6} cancer risk)	Reference Dose[f] mg/kg/day
	Human	Animal/In Vitro				
Terbacil (I)		Increased relative liver weights	E	10-110	90 (lifetime HA)	0.013
Atrazine		Decreased body weight gain	C	<10-1,000	3	0.035
Chlorothalonil (I)		Renal tubular vacuolization	B2	0.04-0.37	500 (DWEL, HA) (1.5)	0.015
Nitrate (as N) (I)	Methemoglobinemia	Reproductive toxicity		Up to 200,000 (I) 500-10,000 (U)	10,000	1.6
Uranium (I)	Nephritis		A	Up to 300	20 (0.7)	
Boron (I)		Testicular atrophy, spermatogenic arrest	D	190-28,000 (I)		0.09
Lindane (U)		Morphological changes of kidney and liver cells	C	0-1	0.2	0.0003
Endrin (U)		Liver lesions (mild); occasional convulsion	D	0-1	2	0.0003
Chlordane (U)		Liver hypertrophy (regional)	B2	--	0.2 (0.03)	0.00006
Methoxyclor (U)		Reproductive effect (litter loss)	D	1-10	40	0.005
Benzene (U)	CNS and bone marrow depression; leukemia, anemia; effects on heart, liver, adrenal gland		A	--	5 (1)	--
Toluene (U)	CNS and bone marrow depression; anemia; effects on heart, liver, adrenal gland	Changes in liver and kidney weights; reproductive effect	D	--	1,000	0.2

1,2-Dichloroethane (U)	Pathological changes in lung, heart, liver, kidney, adrenal gland	B2	--	5 (0.4)	--
Ethylbenzene (U)	Liver and kidney toxicity	D	--	700	0.1
Tetrachloroethylene (U)	Hepatotoxicity (increased relative liver and kidney weights, depressed body weight)		--	5	0.01
Pentachlorophenol (U)	Liver and kidney pathology, feto-maternal toxicity	B2	Up to 100	1 (0.3)	0.03
Polychlorinated biphenyls (PCBs) (U)	Chloracne; lymphoid gland atrophy; immunological effects	B2	--	0.5 (0.005)	
Polyaromatic hydrocarbons (PAHs) (U)					
Benzo(a)pyrene Fluoranthrene	Nephropathy; increased liver weight; hematologic alterations; clinical effects (increased SGPT levels)	B2	--	0.2 (5)	0.04
Phenanthrene Anthracene	No observed effects	D	--	--	0.3
Phthalates, di-n-butyl (U)	Increased mortality		50	--	0.1
Antimony (U)	Gastrointestinal effects	D	25	6	0.0004
Chromium (U)	Renal tubular necrosis	D	1-200(U)	100	0.005
Arsenic (I) (U)	Skin (hyperpigmentation, keratosis); vascular complications; neurotoxicity; liver injury	A	Max. 50 (U) Avg. 1 1-190 (I)	50 (0.000002)[a]	0.0003

TABLE 4.2 (continued)

Chemical[a]	Health Effects[b]		Carcinogen Classification[c]	Concentration[d] μg/l	Maximum Contaminant Level[e] μg/l (10^{-6} cancer risk)	Reference Dose[f] mg/kg/day
	Human	Animal/In Vitro				
Beryllium (U)	Contact dermatitis; pulmonary effects	Skeletal effects; genotoxicity	B2	Max. 50	4 (0.008)	0.005
Cadmium (U)	Pulmonary and renal tubular effects; skeletal changes associated with effects on calcium metabolism	Reproductive/teratogenic effects; effect on myocardium	D	1-15 (U)	5	0.0005
Selenium (I) (U)	Nail changes; Hair loss; Skin lesions; nervous system effects	Reproductive effects; genotoxicity		Up to 300 (I) Max. 100 (U)	50	0.005
Zinc (U)	Gastrointestinal distress; diarrhea	Poor growth	D	200 (can be 1,000 from galvanized metal processing)	--	0.3
Mercury (U)	Nervous system effects; kidney effects	Genotoxicity	D	Max. 1	2	0.0003
Nickel (U)	Contact dermatitis	Reproductive effects; genotoxicity	D	Up to 200 Avg. <100	100	0.005
Cyanide (U)	Nausea, confusion, convulsion, paralysis, coma, cardiac arrhythmia, respiratory stimulation followed by respiratory failure		D	Up to 300	200	0.022

[a] For consistency, these are listed in the order in which they appeared in Table 2.23. I = irrigation return flow and U = urban stormwater. See Table 2.23 and the discussion in Chapter 2 for background information on chemicals in this table.

[b] U.S. EPA, 1992, 1993a, 1993b; Amdur et. al, 1991; Friberg et. al., 1986; Fishbein et. al., 1987 The inclusion of the health effects information is not meant to be comprehensive. Instead, the primary end point of concern, the endpoint on which the drinking water standard is based, and readily available information are presented.

[c] EPA carcinogen classifications (U.S. EPA, 1993a, 1993b). A = sufficient evidence for humans; B1 = limited evidence for humans and sufficient evidence in experimental animals; B2 = inadequate/limited evidence for humans, but sufficient evidence in experimental animals; C = limited evidence in experimental animals with no human data; D = inadequate or no data; E = sufficient evidence of noncarcinogenicity.

[d] See Table 2.23 and the discussion in Chapter 2 for background information on reported concentrations.

[e] Maximum contaminant levels established by EPA as drinking water standards for chemicals. Where no MCL has been established for a chemical, the health advisory (HA) level (lifetime or drinking water equivalent level (DWEL)) is provided. The level in drinking water corresponding to a cancer risk of 1 in 1 million is shown within parentheses. (U.S. EPA, 1993a, 1993b).

[f] Reference dose = estimated doses for a daily exposure (expressed as milligrams of chemical per kilogram of body weight per day) that is likely to be without an appreciable risk of deleterious effects during a lifetime (U.S. EPA, 1993a, 1993b).

[g] OEHHA, 1992. Recommended public health level for arsenic in drinking water. Office of Environmental Health Hazard Assessment, California Environmental Protection Agency, Berkeley, California.

various treated municipal wastewaters are shown in Tables 4.3, 4.4, and 4.5, and compared to MCLs and reference doses. The typical organic pollutants identified in activated sludge secondary effluent (which is the influent to Orange County Water District's Water Factory 21) are shown in Table 4.3, while typical chemicals in secondary-treated municipal wastewater are shown in Table 4.4, and chemicals found in activated-sludge treated secondary effluent from the City of Phoenix's 23rd Avenue Plant are shown in Table 4.5. The chemical concentrations were tabulated from information in Chapter 2. These concentrations and their possible health effects are useful in providing perspective on source waters and the constituents in them that might have a negative impact on human health.

The chemical monitoring data on irrigation return flow and stormwater runoff in Table 4.2 show that the highest concentrations detected exceeded the MCLs for atrazine, nitrate, lindane, uranium, pentachlorophenol, chromium, arsenic, beryllium, and lead. The concentrations for chlorothalonil, arsenic, beryllium, and pentachlorophenol exceeded the levels corresponding to a 1 in 1 million (1×10^{-6}) cancer risk. When the MCLs are exceeded, the reference doses are also exceeded, except for atrazine and beryllium. No MCL has been established for boron, but based on the consumption of 2 liters of water by a 70-kg person at the highest level detected, the reference dose for boron is exceeded.

For the chemicals found in secondary treated municipal wastewaters shown in Table 4.4, the inorganic chemicals boron, cadmium, lead, molybdenum, mercury, and nickel exceeded the drinking water standard or MCLs. Similarly, the organic chemicals in Tables 4.3 and 4.5 were all within limits specified by the MCLs.

Although various inorganic and organic chemicals can clearly be identified in wastewaters that have been or might be used for recharging ground waters, additional treatment prior to recharge will reduce the concentrations of many of these substances, as will soil-aquifer treatment and conventional potable water treatment at the point of extraction. Thus the actual risks to human health will undoubtedly be considerably lower than those implied by the tables. Therefore, the recharge of ground waters with waters of impaired quality is not likely to present unacceptable risks from these inorganic and organic chemical constituents when the extracted water is used for human consumption. However, at the same time the data indicate that some potential source waters have higher concentrations of chemicals of potential health concern than others. Thus, the choice of source water and decisions on monitoring recharged aquifers and the extracted water should take into account the presence of these trace inorganic and organic chemicals and their possible health impacts.

Disinfectants and Disinfection By-Products

One health concern connected with the use of recharge water arises from the use of disinfectants and the formation of disinfection by-products (DBPs), which

TABLE 4.3 Typical Organic Priority Pollutants in Activated Sludge Secondary Effluent from the County Sanitation Districts of Orange County, California

	Secondary Effluent Concentration ($\mu g/l$)	Maximum Contaminant Level[a] ($\mu g/l$ (10^{-6} cancer risk)	Carcinogen Classification[b]	Reference Dose[c] (mg/kg/day)
Chloroform	3.5	100 (0.006)	B2	0.01
Bromodichloromethane	0.46	100 (0.0006)	B2	0.02
Dibromochloromethane	0.71	100 (0.004)	C	0.02
Bromoform	0.46	100 (0.004)	B2	0.02
1,1,1-Trichloroethane	4.8	7 (HA) (0.004)	C	0.03
Trichloroethylene	1.1	5 (0.0003)	B2	—
Tetrachloroethylene	3.6	5	—	0.01
Carbon tetrachloride	0.05	5 (0.0003)	B2	0.0007
Chlorobenzene	0.13	—	—	—
1,3-Dichlorobenzene	0.25	600	D	0.09
1,4-Dichlorobenzene	1.9	600	D	0.09
1,2-Dichlorobenzene	0.74	75	C	0.1
1,2,4-Trichlorobenzene	0.31	70	D	0.1
Naphthalene	0.11	20 (HA)	D	0.004
Ethylbenzene	0.04	700	D	0.1
2,4-Dichlorophenol	0.16	20 (HA)	D	0.003
2,4,6-Trichlorophenol	0.13	— (0.003)	B2	—
Pentachlorophenol	1.23	1 (0.0003)	B2	0.03
PCB (Arochlor 1242)	0.4	0.5 (0.000005)	B2	—
Lindane	0.11	0.2	C	0.0003
DDT	0.01	—	—	—
Di-n-butyl phthalate	0.94	—	—	—
Diethyl phthalate	1.14	5,000 (HA)	D	0.8
Bis(2-ethylhexyl) phthalate	11	—	B2	—
Isophorone	0.3	100 (HA) (0.04)	C	0.2

[a]Maximum contaminant levels established by EPA as drinking water standards for chemicals. Where no MCL has been established for a chemical, the health advisory (HA) level (lifetime or DWEL) is provided. The level in drinking water corresponding to a cancer risk of 1 in 1 million is shown in parentheses (U.S. EPA, 1993b).

[b]EPA carcinogen classification (U.S. EPA, 1993a, 1993b): B2 = inadequate/limited evidence for humans, but sufficient evidence in experimental animals; C = limited evidence in experimental animals with no human data; D = inadequate or no data.

[c]Reference dose = estimated doses for a daily exposure (expressed as milligrams of chemical per kilogram of body weight per day) that is likely to be without an appreciable risk of deleterious effects during a lifetime (U.S. EPA, 1993a and b).

TABLE 4.4 Typical Chemicals in Secondary-Treated Municipal Wastewater

	Concentration (μg/l)	Maximum Contaminant Level[a] μg/L (10^{-6} cancer risk)	Carcinogen Classification[b]	Reference Dose[c] (mg/kg/day)
Nitrate	400-30,000	45,000	—	
Arsenic	5-23	50 (.000002)	A	—
Boron	300-2,500	600 (HA)	D	0.09
Cadmium	5-220	5	D	0.0005
Chromium	1-100	100	D	0.005
Copper	6-53	1,300 (AL)	D	—
Lead	3-350	15 (AL)	B2	—
Molybdenum	1-18	40 (HA)	D	0.005
Mercury	2-10	2	D	0.0003
Nickel	3-600	100	D	0.02
Zinc	4-350	2,000 (HA)	D	0.3

[a]Maximum contaminant levels established by EPA as drinking water standards for chemicals. AL = Action Level. Where no MCL has been established for a chemical, the health advisory (HA) level (lifetime or DWEL) is provided. The level in drinking water corresponding to a cancer risk of 1 in 1 million is shown in parentheses (U.S. EPA, 1 993b).

[b]EPA carcinogen classification (U.S. EPA, 1993a, 1993b): B2 = inadequate/limited evidence for humans, but sufficient evidence in experimental animals; D = inadequate or no data.

[c]Reference dose = estimated doses for a daily exposure (expressed as milligrams of chemical per kilogram of body weight per day) that is likely to be without an appreciable risk of deleterious effects during a lifetime (U.S. EPA, 1993a and b).

can result in chemical-related cancer. The various types of disinfectants that currently are used and their by-products are described in Chapter 2. An overview of the health effects of these disinfectants and by-products is discussed in detail in Bull and Kopfler (1991), and the estimated carcinogenic risks of these chemicals are summarized in Table 4.6.

Trihalomethanes (THMs) are one of the by-products of chlorination. The major THMs are chloroform, bromodichloromethane, chlorodibromomethane, and bromoform. These have been evaluated by International Agency for Research on Cancer (IARC) (1987, 1991) and U. S. Environmental Protection Agency (EPA) (1986) as having varying degrees of evidence of carcinogenicity in animals and humans.

Disinfectants or DBPs that have been determined by IARC or EPA to be carcinogenic in animals are 2,4,6-trichlorophenol, formaldehyde, acetaldehyde, bromate (potassium), and bromoform (International Agency for Research on Cancer, 1987, 1991; U.S. Environmental Protection Agency, 1986). Dichloroacetic acid and trichloroacetic acid have been shown to induce liver tumors in

B6C3F1 mice (Parnell et al., 1986; Herren-Freund et al., 1987). In addition, five mutagenic derivatives of a trichlorinated hydroxyfuranone (MX) have been identified as chlorinated by-products. These have been shown to be extremely potent mutagens in the Ames assay, although they are present in small concentrations. MX has been named 3-chloro-4-(dichloromethyl)-5-hydroxy-2 named (5H)-furanone since the compound forms a furanone ring at pH below 5.3. However, at the pH of drinking water and under the neutral conditions of the Ames assay, MX exists in a ring-opened form. In referring to chemistry in neutral water solutions and mutagenicity of MX, MX should be regarded as an oxobutenoic acid, or (z)-2-chloro-3-(dichloromethyl)-4-oxobutenoic acid) (Kronberg et al., 1991).

MICROORGANISMS OF CONCERN

Hundreds of different types of pathogenic microorganisms (i.e., bacteria, viruses, and parasites) are excreted in the fecal material of infected hosts and these can find their way into municipal wastewater (Rao and Melnick, 1986; Straub et.al., 1993). The number and types of pathogenic microorganisms present in wastewater vary by location and over time at a given location. A variety of factors influence the pathogen content of wastewater, including the incidence of disease in the population producing the wastewater, the season of year, the economic status of the population, water use patterns, and the quality of the potable water supply (Rose and Carnahan, 1992). Diseases caused by waterborne microorganisms range from mild gastroenteritis to severe illnesses such as infectious hepatitis, cholera, typhoid, and meningitis. Because of the great variety of factors involved, risk assessments conducted for microorganisms are complicated; a discussion of the issues involved in such assessments follows.

Hazard Identification

The microorganisms of concern when using wastewater to artificially recharge ground water can be identified using data available from past waterborne disease outbreaks. Table 4.7 lists the microorganisms that have been identified as causative agents of waterborne disease in the United States from 1971 through 1990. These outbreaks were associated with a variety of types of water systems—large systems, community systems, small systems, and individual systems—not necessarily parallel to the large systems commonly associated with recharge operations. This information is helpful, however, in identifying the possible range of the microbial hazard. The most commonly identified causative agents were *Giardia*, chemical poisoning, and *Shigella* species. *Giardia lamblia* caused over 18 percent of the illness associated with waterborne disease outbreaks. Enteric viruses (viral gastroenteritis and hepatitis A) were identified as the causative agents of disease in 8.7 percent of the outbreaks during this period.

TABLE 4.5 Typical Chemicals in Activated-Sludge-Treated Secondary Effluent from the City of Phoenix's 23rd Avenue Treatment Plant

	Geometric Mean Concentration of Secondary Effluent (µg/L)		Maximum Contaminant Level[a] (10⁻⁶ cancer risk) (µg/l)	Carcinogen[b] Classification	Reference Dose[c] (mg/kg/day)
	Without Chlorination (27 samples)	With Chlorination (27 samples)			
Aliphatic hydrocarbons					
5-(2-Methylpropyl) nonanes	0.35	0.57	—	—	—
2,2,3,-Trimethylhexane	0.11	0.18	—	—	—
6-Methyl-5-nonen-4-one	0.41	0.94	—	—	—
2,2,3-Trimethylnonane	0.21	0.25	—	—	—
2,3,7-Trimethyloctane	0.12	0.27	—	—	—
Aromatic hydrocarbons					
o-Xylene	0.45	0.5	10,000	D	2
m-Xylene	0.76	1	10,000	D	2
p-Xylene	0.17	0.12	10,000	D	2
C3-benzene isomer	0.56	0.34	—	—	—
C3-benzene isomer	0.48	0.53	—	—	—
Styrene	0.26	0.58	100	C	—
1,2,4-Trimethylbenzene	0.8	1.04	—	—	—
Ethylbenzene	0.19	0.15	700	D	0.1
Naphthalene	0.22	0.63	20 (HA)	D	0.004
Phenanthrene	0.1	0.1	—	—	—
Diethyl phthalate	19	10	5,000 (HA)	D	0.8

Chlorinated aliphatic hydrocarbons				
Chloroform	2.72	3.46 (0.006)	100	B2	0.01
1,1,1-Trichloroethane	2.94	1.41 (0.001)	7 (HA)	C	0.03
Carbon tetrachloride	0.12	0.12 (0.0003)	5	B2	0.0007
Bromodichloromethane	Not detected	0.26 (0.0006)	100	B2	0.02
Trichloroethylene	0.91	0.39 (0.003)	5	B2	—
Dibromochloromethane	Not detected	0.23 (0.004)	100	C	0.02
Tetrachloromethane	2.63	1.69 (0.003)	5	B2	—
Bromoform	Not detected	0.08 (0.004)	100	B2	0.02
Chlorinated aromatics				
o-Dichlorobenzene	3.52	2.4	600	D	0.09
m-Dichlobenzene	0.79	0.38	600	D	0.09
p-Dichlobenzene	2.25	1.81	75	C	0.1
1,2,4-Trichlorobenzene	0.19	0.38	70	D	0.01
Trichlorophenol	0.01	0.02	— (0.003)	B2	—
Pentachlorophenol	0.02	0.04	1 (0 .0003)	B2	0.03
Pentachloroanisole	0.43	0.18	—	—	—

[a]Maximum contaminant levels established by EPA as drinking water standards for chemicals. AL = Action Level. Where no MCL has been established for a chemical, the health advisory (HA) level (lifetime) is provided. The level in drinking water corresponding to a cancer risk of 1 in 1 million is shown in parentheses (U.S. EPA, 1993b).

[b]EPA carcinogen classification (U.S. EPA, 1993a, 1993b): B2 = inadequate/limited evidence for humans, but sufficient evidence in experimental animals; D = inadequate or no data.

[c]Reference dose = estimated doses for a daily exposure (expressed as milligrams of chemical per kilogram of body weight per day) that is likely to be without an appreciable risk of deleterious effects during a lifetime (U.S. EPA, 1993a and b).

TABLE 4.6 Estimated Carcinogenic Risks from By-Products of Various Disinfectants[a]

	HOCl/OCl⁻ ($\times 10^{-6}$)	ClNH$_2$[b] ($\times 10^{-6}$)	ClO$_2$ ($\times 10^{-6}$)	O$_3$ ($\times 10^{-6}$)
Chloroform	0.24	0.48	No Data	0
Bromodichloromethane	0.91	0.18		
Chlorodibromomethane	0.29	0.058		
Bromoform	0.054	0.011		0.025
Dichloroacetic acid	0.00034	0.000068		
Trichloroacetic acid	54	11		
Chloropicrin	0.0016	0.00032		
2,4,6-Trichlorophenol	0.017	0.0034		
Formaldehyde	0.54	0.11		3.0
Hydrogen peroxide				10
Bromate				50
Projected mean risk	56	11		63

[a]Estimates of carcinogenic risks associated with chloramine by-products were made by assuming that chlorinated by-products would form at 20 percent of the level observed with chlorine in the same supply based on the results of Amy et al., 1990.

[b]The maximum likelihood estimates of mean carcinogenic risks have been calculated using data from animal studies described in Bull and Kopfler (1991) and estimated mean concentrations of the byproducts in supplies in which the indicated disinfectants are used. The multistage model (using the TOX-RISK program developed by Clement Associates, Inc.) was used to calculate the extra probability that an individual would contract cancer from drinking the mean concentration of the indicated byproduct for a lifetime. The projected mean risk for each disinfectant was calculated by summing risks estimated from the mean concentrations of the carcinogenic by-products produced by that disinfectant and is expressed in extra cases per million population per lifetime. Source: Bull and Kopfler, 1991. Reprinted from *Health Effects of Disinfectants and Disinfection By-Products*, by permission. Copyright 1991, American Water Works Association.

In the 1980s, use of un-disinfected or inadequately disinfected ground water in general was responsible for 44 percent of the waterborne disease outbreaks that occurred in the United States (Craun, 1991). For outbreaks that occurred because of the consumption of contaminated, un-disinfected ground water from 1971 to 1985, sewage was most often identified as the contamination source. In ground water systems, causative agents were identified in only 38 percent of the outbreaks, with *Shigella* species and hepatitis A virus being the most commonly identified pathogens (Craun, 1990).

At the present time, we cannot completely identify the microbial hazard because of our inability to identify causative agents in approximately one-half of the waterborne disease outbreaks in this country. In these outbreaks, the illness was simply listed as gastroenteritis of unknown etiology. However, retrospective serological studies of outbreaks of acute nonbacterial gastroenteritis from

TABLE 4.7 Causative Agents of Waterborne Disease Outbreaks, 1971 to 1990

	Outbreaks		Illness	
	Number of Cases	Percentage of Total	Number of Cases	Percentage of Total
Gastroenteritis, unknown cause	293	49.66	67,367	47.26
Giardiasis	110	18.64	26,531	18.61
Chemical poisoning	55	9.32	3,877	2.72
Shigellosis	40	6.78	8,806	6.18
Viral gastroenteritis	27	4.58	12,699	8.91
Hepatitis A	25	4.24	762	< 1
Salmonellosis	12	2.03	2,370	1.66
Campylobacterosis	12	2.03	5,233	3.67
Typhoid fever	5	< 1	282	< 1
Yersiniosis	2	< 1	103	< 1
Crytosporidiosis	2	< 1	13,117	9.20
Chronic gastroenteritis	1	< 1	72	< 1
Toxigenic *E. coli*	2	< 1	1,243	< 1
Cholera	1	< 1	17	< 1
Dermatitis	1	< 1	31	< 1
Amebiasis	1	< 1	4	< 1
Cyanobacteria-like bodies	1	< 1	21	< 1
Total	564	100	138,247	100

Source: Craun, 1991; Herwaldt et al., 1992.

1976 through 1980 indicated that 42 percent of outbreaks where no causative agent was identified probably were caused by the Norwalk virus (Kaplan et al., 1982). Thus, the Norwalk virus may be responsible for approximately 23 percent of all reported waterborne disease outbreaks in the United States (Keswick et al., 1985).

The difficulty in the isolation of many enteric viruses from clinical and environmental samples probably accounts for the limited number of viruses identified as causes of waterborne disease. For example, there are no standardized, routine procedures available for isolating and identifying hepatitis A and E viruses in environmental (i.e., soil and water) samples. The ability to analyze samples for *Cryptosporidium* is restricted to only a few laboratories. There are no methods available for culturing the Norwalk virus in the laboratory. As methods for the detection of enteric viruses and parasites have improved, the percentage of waterborne disease identified as having a viral or parasitic etiology has increased.

In addition to the hazard of acute microbial disease that results in waterborne disease outbreaks, the level of endemic microbial disease associated with drinking water must be identified. In the single epidemiological study that has been conducted to determine the contribution of drinking water to endemic gastrointestinal illness, it was found that approximately one-third of the outbreaks could be the result of consuming treated drinking water that met all water quality standards and contained no pathogens detectable by current technologies (Payment et al., 1991).

It has been recognized recently that exposure to microbial pathogens in drinking water may also lead to chronic health problems such as diabetes (Gerba and Rose, 1993). This association must be further investigated so that the microbial hazard can be identified accurately.

Dose-Response Assessment

Three different responses to microbial exposure are possible: infection that remains subclinical (i.e., inapparent), infection that results in clinical illness, and infection that leads to illness and subsequent death. The response used in the dose-response determination depends on the purpose of the risk assessment. For example, in balancing the risks between pathogenic microorganisms and DBPs in water, it might be desirable to use mortality as the end point for both cases. However, other end points such as number of life-years lost or "quality adjusted" life years lost may be used (Putnam and Graham, 1993). In the latter two cases, illness and mortality could be used in the determination.

Dose-response data on the ability of a microorganism to cause infection are generally obtained by exposing a group of animals or human volunteers to different doses of the microorganisms of interest. In the case of enteric viruses, human volunteers must be used because animals are not infected by the viruses of interest. Infection is determined directly by detecting the microorganisms in the fecal material or other bodily fluids, or indirectly by detecting an antibody response to the microorganism. The infective dose of several enteric microorganisms is shown in Table 4.8.

These experimentally obtained dose-response data have been analyzed to determine whether mathematical models can be used to describe them. Several models are available for describing dose-response relationships. Two models, the simple exponential and the modified exponential (ß), have generally been used to model the response of humans to enteric pathogens (Haas, 1983; Regli et al., 1991). The experimental data are fit to a particular model, enabling one to determine the value of the constants in the model. Once the values of the constants are known, the probability of infection from ingestion of any number of microorganisms can be calculated.

When morbidity or mortality are the endpoints of interest, data are generally obtained from medical and hospital records. The relationships between infection

TABLE 4.8 Values Used to Calculate Risks of Infection, Illness, and Mortality from Selected Enteric Microorganisms

	Probability of Infection from Exposure to One Organism (per million)	Ratio of Clinical Illness to Infection (%)	Mortality Rate (%)	Secondary Spread (%)
Campylobacter	7,000			
Salmonella typhi	380			
Shigella	1,000			
Vibrio cholerae	7			
Coxsackieviruses		5-96	0.12-0.94	76
Echoviruses	17,000	50	0.27-0.29	40
Hepatitis A virus		75	0.6	78
Norwalk virus			0.0001	30
Poliovirus 1	14,900	0.1-1	0.9	90
Poliovirus 3	31.000			
Rotavirus	310,000	28-60	0.01-0.12	
Giardia lamblia	19,800			

Source: Rose and Gerba, 1991; Gerba and Rose, 1993.

and clinical illness for several enteric viruses are shown in Table 4.8. Mortality rates for enteric viruses are also given in Table 4.8.

Dose-response data are generally obtained from studies of relatively small groups of healthy volunteers; thus they represent average or possibly best-case situations. Certain populations may be more at risk from exposure to a given dose of pathogens than others. For example, very young and very old individuals have a higher risk of severe illness and even death from exposure to pathogens than do other population groups. Individuals with suppressed immune systems may also be more susceptible to infection, illness, and death than healthy individuals. It has been estimated that 17 percent of the U.S. population may be classified as "at increased risk" for the purposes of risk calculations (C. P. Gerba, personal communication to M. Yates, 1993).

When assessing microbial risk, it is necessary to consider secondary spread. This phenomenon occurs when an individual who has been infected by consuming water containing pathogenic microorganisms transmits the infection to another individual. Secondary spread can be significant for some enteric microorganisms, as shown in Table 4.8. This phenomenon is specific to microorganisms and has no parallel in the assessment of risks from chemicals in water.

Another issue that has not been adequately addressed in the dose-response determination or the exposure assessment is that of aggregated (i.e., a group of several microorganisms stuck to one another) or solids-associated microorganisms. Most dose-response studies are performed using laboratory-grown and purified, monodispersed pathogens. However, in the environment most enteric viruses occur as aggregated units or associated with cellular debris (Sobsey et al., 1991). During sample analysis, a virus aggregate comprising of several tens to hundreds of viral particles may be counted as only one infectious unit. Thus, the exposure to pathogens in drinking water may be higher than is reflected by analysis of a contaminated water sample.

Exposure Assessment

Assessing the exposure of an individual to pathogens in recovered water is the most difficult and uncertain aspect of the risk assessment process. Exposure to pathogens may occur by direct ingestion of or contact with the recharge water at the surface if recharge is by infiltration, or it may occur by ingestion of recovered water that has been contaminated by the recharge process. The level of exposure as well as the pathogens of concern depend on the route of exposure. For example, exposure to bacteria, viruses, and parasites is likely if reclaimed water in an infiltration basin is directly ingested. But because bacteria and parasites are generally removed to a greater extent than enteric viruses during infiltration through soil, viruses are of greater concern when dealing with exposure to recovered recharge water.

The number and types of pathogens in recovered water depend on the level of treatment the water has received. As stated earlier, the concentrations of different pathogens in raw wastewater vary among communities and over time within a given community. For instance, reported concentrations of microorganisms in raw wastewater range from zero to several hundred thousand *Giardia* cysts per liter, hundreds of thousands of viral particles per liter, and several tens of thousands of bacterial pathogens per liter.

Past studies on the efficacy of various wastewater treatment processes on pathogen inactivation have focused on the removal of indicator microorganisms such as total and fecal coliform bacteria, and occasionally enteroviruses such as poliovirus. In one such study, data on the efficiency of removal of enteroviruses at water reclamation plants in southern California collected during the 10-year period from 1979 to 1989 were summarized (Yanko, 1993). Table 4.9 shows the average concentration of enteroviruses detected in unchlorinated effluents (primary, secondary, and tertiary), as well as chlorinated final effluent. During the 10-year study period, a total volume of 613,639 l of final effluent in the form of 590 samples was analyzed. Of these, only one sample was found to contain an enterovirus, coxsackievirus B3. It is important to note that the efficiency of the methods used to collect and recover the viruses averaged 41 percent.

There is little, if any, information on the concentration of several pathogens of public health significance including hepatitis A and E viruses, rotavirus, Norwalk virus, *Giardia*, and *Cryptosporidium*. A study in Florida examined the

TABLE 4.9 Removal of Viruses at Six Los Angeles County, California, Treatment Plants

	Primary Effluent		Secondary Effluent		Tertiary Effluent		Final Effluent	
	pfu	No. Positive/ No. Sampled	pfu	No. Positive/ No. Sampled	pfu	No. Positive/ No. Sampled	Total Volume Sampled (l)	No. positive/ No. sampled
Long Beach	NR	NR	NR	NR	NR	NR	74,591	0/84
Los Coyotes	NR	NR	NR	NR	NR	NR	67,668	0/74
Pomona	23,000	13/13	650	NR	57	6/12	147,774	0/124
San Jose Creek	160,000	8/8	55	NR	NR	NR	139,390	0/130
Whittier Narrows	76,000	8/8	56	NR	NR	NR	131,986	0/124
Valencia	NR	NR	NR	NR	NR	NR	52,229	1/54

Note: NR = not reported. Pfu = Plaque-forming units per 387 liters.

Source: Yanko, 1993.

removal rates of enteroviruses, indicator bacteria and viruses, *Giardia*, *Cryptosporidium*, and helminths in a full-scale operating wastewater treatment plant (Rose and Carnahan, 1992). Average removal for each of the organisms measured at each stage of the treatment process over a 1-year period are shown in Table 4.10.

The susceptibility of different pathogens to various disinfection processes has also been measured. Most of the information on disinfection efficiency, especially for processes other than those involving chlorination, has been obtained from drinking water treatment rather than wastewater treatment. It is uncertain whether the trends found in drinking water studies will be consistent in wastewater disinfection studies. A summary of the efficacy of various disinfection processes on removal of pathogens and indicator bacteria from drinking water is shown in Table 4.11.

To determine the exposure of an individual to pathogens in recovered water, the ingested number of each of the pathogens of concern must be known. For direct ingestion of recovered water, the only report on this topic used an exposure of 100 ml (Rose and Carnahan, 1992). The concentrations of the pathogens in this case would be the concentrations remaining after treatment of the water. In the case of ingestion of recovered water, an ingestion of 2 liters of water per day is usually assumed. The concentration of microorganisms in the ground water is more difficult to determine because it is dependent on the level of treatment prior to recharge and the removal of the various microorganisms during infiltration through the soil and aquifer; these removal rates are normally unknown.

Risk Characterization

Using the data from Table 4.8, and assuming an exposure of 2 liters per day of water containing a known concentration of pathogens, the probability of infection, illness, and death from exposure to a given microorganism can be calculated. Regli et al., (1991) compared the simple exponential and modified exponential (ß) with experimental dose-response data. They found that the ß model fit the echovirus 12, poliovirus 3, and rotavirus exposure data best; whereas the exponential model fit the poliovirus 1 data best. Using three different concentrations that might be found in drinking water, the annual risks of infection, disease, and mortality from exposure to hepatitis A virus and rotavirus in 2 liters of drinking water per day are shown in Figures 4.1 and 4.2, respectively.

The data on pathogen concentrations in the final effluent from one of the St. Petersburg, Florida, water reclamation plants were used to calculate the risks associated with accidental ingestion of 100 ml of the water (Rose and Carnahan, 1992). The following assumptions were used in the risk estimation: (1) the echovirus and rotavirus models were used to reflect a moderately infective and a highly infective virus (as demonstrated by dose-response data); (2) the *Giardia*

TABLE 4.10 Average Removal of Pathogen and Indicator Microorganisms in a Wastewater Treatment Plant, St. Petersburg, Florida

	Raw Wastewater to Secondary Wastewater		Secondary Wastewater to Postfiltration		Postfiltration to Post-Disinfection		Post-Disinfection to Poststorage		Raw Wastewater to Poststorage	
	Percentage	$\log_{10}$	Percentage	$\log_{10}$	Percentage	$\log_{10}$	Percentage	$\log_{10}$	Percentage	$\log_{10}$
Total coliforms (cfu/100 ml)	98.3	1.75	69.3	0.51	99.99	4.23	75.4	0.61	99.999992	7.10
Fecal coliforms (cfu/100 ml)	99.1	2.06	10.5	0.05	99.998	4.95	56.8	0.36	99.999996	7.42
Heterotrophic plate count bacteria (cfu/ml)	99.1	2.06	81.1	0.72	99.98	3.77	none		99.99996	6.55
Coliphage 15597 (pfu/ml)	82.1	0.75	99.98	3.81	90.5	1.03	90.3	1.03	99.99997	6.61
Coliphage C (pfu/ml)	none		99.94	3.20	99.7	2.49	71.0	0.54	99.99991	6.23
Enterovirus (pfu/100 1)	98.0	1.71	84.0	0.81	96.5	1.45	90.91	1.04	99.999	5.01
Giardia (cysts/100 1)	93.0	1.19	99.0	2.00	78.0	0.65	49.5	0.30	99.993	4.13
Cryptosporidium (oocysts/100 1)	92.8	1.14	97.9	1.68	61.1	0.41	8.5	0.04	99.95	3.26
Helminths (ova/l)	>75.0								>99.6	

Note: cfu = colony-forming unit; pfu = plaque-forming unit. Source: Rose and Carnahan, 1992.

TABLE 4.11 Inactivation of Indicator and Pathogenic Microorganisms by Various Disinfectants

	Free Chlorine (4-5 C, pH 6-8)		Chloramine (5 C, pH 7-9)		Chlorine Dioxide (5 C, pH 6-7)		Ozone (4-5 C, pH 6-8)		Ultraviolet Light (variable)	
	C·T	Inactivation (%)	C·T	Inactivation (%)	C·T	Inactivation (%)	C·T	Inactivation (%)	mW-s cm^{-2}	Inactivation (%)
E coli	2.5	99.9	113	99	0.48	99	0.006-0.02	99	6.5	99.9
Poliovirus 1	1.1-2.5	99	1,420	99	0.2-6.7	99	0.2	99	21	99.9
Rotavirus SA11	0.03	99.9	4,034	99	0.2-0.3	99	0.019-0.064	99.9	25	99.9
Human rotavirus	0.03	99.9	ND	ND	ND	ND	ND	ND	0.006-0.036	99.9
Hepatitis A virus	1.8	99.99	592	99.99	1.7	99	ND	ND	ND	ND
MS2 coliphage	0.25	99.99	2,100	99.99	5.1	99	ND	ND	ND	ND
Giardia lamblia	90-170	90	ND	ND	ND	ND	0.53	99	100	99.9
Giardia murls	>150	90	1,400	99	10.7	99	1.94	99	ND	ND
Cryptosporidium	>>1080	<95	ND	ND	ND	ND	ND	ND	ND	ND

Note: ND = no data. C·T = product of the disinfectant in milligrams per liter and contact time in minutes for 99 percent inactivation.

Source: Adapted from Bull et al., 1990; Sobsey, 1989.

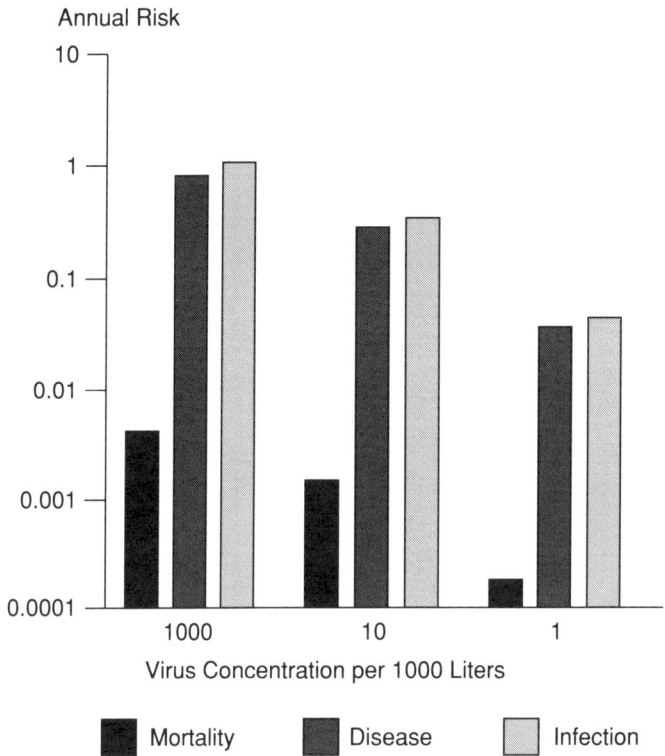

FIGURE 4.1 Annual risk infection, disease, and mortality for hepatitis A virus in drinking water.

model was used for both *Giardia* and *Cryptosporidium* because there is no model for *Cryptosporidium*; (3) all cysts and oocysts were assumed to be viable and infective; and (4) concentrations in the final product (per 100 l) were calculated to levels per 100 ml and used as exposures. The calculated risks are shown in Table 4.12. The risks ranged from a high of 1.1 in 10,000 for *Cryptosporidium* to a low of 2 per 100 million for viruses.

Asano et al. (1992) used the ß model to calculate the risk associated with exposure to viruses in ground water recharge operations. They assumed that the nearest domestic well to a recharge site could draw water that contains 50 percent recovered wastewater that has been underground for 6 months after percolating through 3 m (9.8 ft) of unsaturated soil. The rate of virus removal/inactivation was assumed to be 0.69 per day. They also assumed that an individual consumes 2 liters of water per day for 70 years. Results of their calculations for poliovirus 1, echovirus 12, and poliovirus 3 are presented in Table 4.13. The equation used to calculate the fraction of viruses remaining after infiltration through the soil must also be carefully examined to determine its applicability to

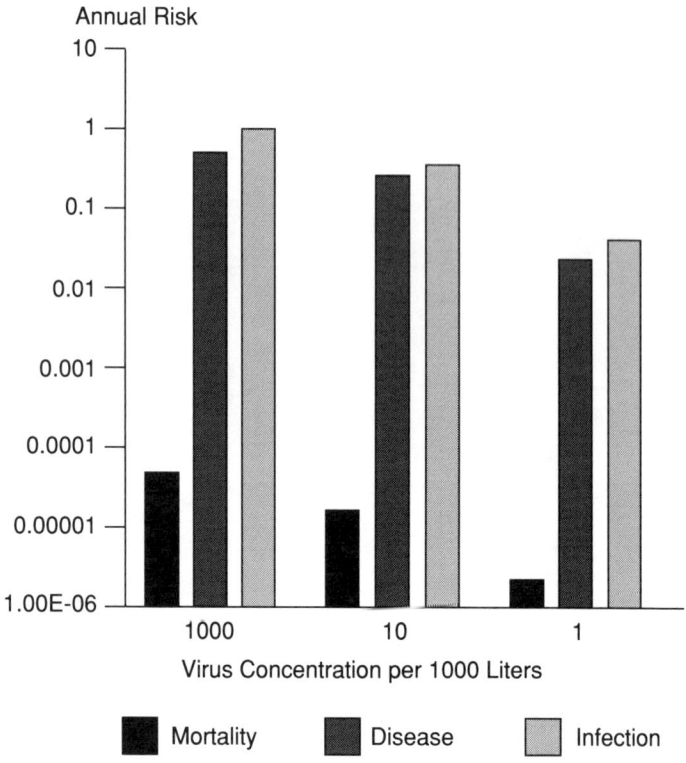

FIGURE 4.2 Annual risk infection, disease, and mortality for rotavirus in drinking water.

a variety of soil types, pathogens, and hydrogeologic and environmental conditions.

An effort to calculate infection risks associated with exposure to recovered water produced by California treatment plants was conducted recently (Tanaka et al., 1993). In this risk assessment, measured concentrations of enteroviruses in unchlorinated secondary effluent, rather than final effluent (in which viral concentrations are generally below detection limits), were used as the starting point. Full tertiary treatment (consisting of coagulation, flocculation, sedimentation, filtration, and disinfection) would be expected to reduce the viral concentration by 5.2 logs from the secondary effluent. If the secondary effluent was directly chlorinated, the virus concentration would be reduced by 3.9 logs. The researchers assumed that 2 l ground water containing 50 percent reclaimed water was consumed per day. They also assumed that the rate of virus inactivation/removal during transport through 3 m (10 ft) unsaturated zone and 6 months retention in the aquifer was 0.1 day^{-1}. The annual risks of virus infection from

TABLE 4.12 Probability of Infection from Accidental Ingestion of 100 ml of Recovered Water Containing Viruses and Protozoa

Levels in Treated Water (per 100 l)	Exposure per 100 ml	Estimated Risk of Infection in Exposed Population		
		Rotavirus Model	Echovirus Model	*Giardia* Model
Viruses				
0.01 plaque-forming units	1.0×10^{-5}	6.2×10^{-6}	2.0×10^{-3}	
0.13 plaque-forming units	1.3×10^{-4}	6.0×10^{-5}	2.7×10^{-7}	
Giardia				
0.49 cysts	4.9×10^{-4}			9.8×10^{-6}
0.89 cysts	8.9×10^{-4}			1.88×10^{-5}
1.67 cysts	1.77×10^{-3}			3.3×10^{-5}
3.3 cysts	$3.3 \times \times 10^{-3}$			6.6×10^{-5}
Cryptosporidium				
0.75 oocysts	7.5×10^{-4}			1.5×10^{-5}
5.35 oocysts	5.35×10^{-3}			1.1×10^{-4}

Source: Rose and Carnahan, 1993.

TABLE 4.13 Risk of Contracting at Least One Infection from Exposure to Viruses in Ground Water Recharged with Chlorinated Tertiary Reclaimed Wastewater

	Maximum Concentration	
Viruses	111 plaque-forming units 100 l^{-1}	1 plaque-forming units 100^{-1}
Lifetime Risk		
Echovirus 12	4.13E-6	3.72E-8
Poliovirus 1	3.78E-7	3.40E-9
Poliovirus 3	1.59E-6	1.43E-8
Annual Risk		
Echovirus 12	5.90E-8	5.31E-10
Poliovirus 1	5.40E-9	4.86E-11
Poliovirus 3	2.27E-8	2.04E-10
Daily Risk		
Echovirus 12	1.62E-10	1.46E-12
Poliovirus 1	1.48E-11	1.33E-13
Poliovirus 3	6.21E-11	5.60E-13

Source: Asano et al., 1992.

exposure to ground water recharged with reclaimed water are shown in Table 4.14. Calculated infection risks for effluent treated beyond unchlorinated secondary ranged from 8 per 100 million to 5.8 per 100 billion.

RISKS FROM DISINFECTANTS AND DISINFECTION BY-PRODUCTS VERSUS RISKS FROM PATHOGENS

From a public health perspective, a central issue associated with the artificial recharge of ground water is how to balance the risks associated with pathogenic microorganisms against those from disinfection by-products (DBPs). To develop an optimal strategy for the protection of public health, the relative risks associated with various concentrations of pathogenic microorganisms and DBPs must be weighed in light of the benefits brought by use of an effective disinfectant. The carcinogenic risk from the DBPs should be compared to the risk of infection from pathogens. Such a comparison grossly oversimplifies the risk-benefit ratio, however, and should serve simply as a crude comparison to obtain more information for consideration rather than an accurate quantitative assessment.

The carcinogenic risks from disinfectants and DBPs have been estimated by Bull and Kopfler (1991). Bromodichloromethane presents the highest risks based on adenocarcinomas of the large intestine. The risks estimated from the chlori-

TABLE 4.14 Predicted Annual Virus Infection Risk from Exposure to Ground Water Recharged with Reclaimed Water (Monte Carlo Simulation, n=500)

Tertiary Treatment	Treatment System			
	OCSD TF	OCSD AS	Pomona AS	MRWPCA AS
Full (5.2 long inactivation/removal)	8.32E-10	5.81E-11	1.69E-10	3.25E-9
Direct chlorination of secondary effluent (3.9 log inactivation/removal)	2.09E-08	1.46E-09	4.25E-09	8.17E-08
Unchlorinated secondary effluent (0 log inactivation/removal)	1.32E-4	1.32E-6	2.68E-5	5.1 5E-4

Source: Tanaka et al., 1993.

nated acetic acid are even higher when mouse liver tumors are considered. The chemicals have not been tested in a second species.

As a crude comparison, Bull et al., (1990) indicated that the probability of mortality induced by improperly disinfected drinking water would exceed the carcinogenic risks introduced by chlorine by as much as 1,000-fold. If disinfection were removed from a contaminated water system that depends on such disinfection, everyone in the community would contract one or more waterborne diseases during their lifetime. Theoretically, the number of deaths from such diseases would equal the mortality rate for the same diseases. The mortality rates for various enteric bacteria and enteroviruses range from 0.12 to 0.94 percent. The carcinogenic risks from chloroform at the mean concentration found (26.4 µg/l) has been estimated to range from 0.24 to 3.0×10^{-6} based on the use of MLE or UCL 95 percent and different tumor sites (Bull and Kopfler, 1991). The risk estimates for other chemicals considered individually range from 0.00034 to 140×10^{-6}. For pathogens, the estimated lifetime risk of mortality from exposure to echovirus 6 at 1 viral infections unit per 1,000 liters of water is greater than 10^{-5}. At one infectious unit of hepatitis in 1,000 liters, the estimated annual risk of mortality is greater than 10^{-4}. The actual risk varies depending on the presence of the number of pathogens. In domestic sewage and polluted surface water (streams) in the United States, the concentrations of enteric organisms (enteric virus, *Cryptosporidium, Giardia*) have ranged from 0.1 to 10 per 100 ml (Bull et al., 1990). Coliforms were estimated at 10^{-5} and 10^{-9} per 100 ml. If the estimated concentration range for enteric virus is extrapolated to echovirus 6 and the lowest concentration is assumed (1 per 100 ml in polluted stream), then the concentration of echovirus 6 for 1,000 liters of water would be 1,000,000 rather than 1 infectious unit, and the resulting estimated lifetime risk of mortality would be significantly higher than 10^{-4}.

HEALTH IMPLICATIONS FROM NONPOTABLE USES

The public health implications of nonpotable reuse, except where intended for market crops to be eaten raw, have not been addressed as extensively as the implications of potable reuse because nonpotable reuse has been practiced widely for decades without public concern and because the exposure from nonpotable reuse, and thus the risk, is more limited. An upcoming study from the National Research Council, "Use of Treated Municipal Wastewater Effluent and Sludge in Crop Production," will address issues associated with using wastewater and sludge in food crop production in depth; the report is expected to be available in early 1995. In general, however, the health implications of using recovered ground water for nonpotable purposes are more easily addressed than those associated with potable reuse. For most nonpotable uses of water, pathogenic microorganisms represent the major health concern. However, chemical constituents may represent a health hazard for some uses, such as crop irrigation and aquaculture, where inorganic or organic constituents of health significance may accumulate in crops or fish.

Recommended water quality criteria and parameters have been developed for several nonpotable water applications, such as irrigation water and industrial cooling and process water (U.S. Environmental Protection Agency, 1973, 1992; Ayers and Westcott, 1985). In addition, many states have developed comprehensive water reclamation and reuse criteria for nonpotable uses of treated municipal wastewater (State of California, 1978; Florida Department of Environmental Regulations, 1990; State of Texas, 1990; Pawlowski, 1992; State of Washington, 1993); hence, state water reuse criteria, which are principally directed at public health protection from pathogenic microorganisms, often are applicable to extracted ground water containing previously recharged wastewater. EPA's *Guidelines for Water Reuse* (U.S. Environmental Protection Agency, 1992) contain recommended wastewater treatment processes and water quality limits for a wide variety of nonpotable types of reuse.

Pathogens are a concern where there is human exposure to the water (by contact, inhalation, or ingestion) or to food or other objects that have come in contact with the water. The types and concentrations of pathogens that may be present in source water used for recharge are presented in Chapter 2. Protection of public health is achieved by (1) reducing the concentrations of pathogenic bacteria, parasites, and enteric viruses in the source water prior to recharge; (2) disinfecting the water upon extraction, if necessary; or (3) limiting public or worker exposure to the water or fomites (objects that might be contaminated with infectious agents). While many bacteria and most, if not all, larger organisms generally are effectively removed after percolation through a short distance of the soil mantle at percolation sites, viruses have been isolated by several investigators examining a variety of recharge operations (Gerba and Goyal, 1985).

Under favorable conditions, pathogens can survive for long periods on crops or in water or soil. Factors that affect survival include the number and type of organism, soil organic matter content (presence of organic matter aiding survival), temperature (longer survival at low temperatures), humidity (longer survival at high humidity), pH, amount of rainfall, amount of sunlight (solar radiation being detrimental to survival), protection provided by foliage, and competitive microbial fauna and flora. Survival times for any particular microorganism exhibit wide fluctuations under differing conditions (Feachem et al., 1983).

It has been reported that viruses and other pathogens in irrigation water do not readily penetrate fruits or vegetables unless the skin is broken (Bryan, 1974). In a study in which soil was inoculated with poliovirus, viruses were detected in the leaves of plants only when the plant roots were damaged or cut (Shuval, 1978). Although adsorption of viruses by plant roots, and subsequent translocation to stems and leaves, has been reported (Murphy and Syverton, 1958), it probably does not occur with sufficient regularity to be a mechanism for the transmission of viruses. Therefore, the likelihood that pathogens would be translocated through trees or vines to the edible part of crops is extremely low.

Viruses and many pathogenic bacteria are in the respirable size range; hence, a possible direct means of human infection is by inhalation of aerosols containing pathogenic microorganisms. Infection or disease also may be contracted indirectly by aerosols deposited on surfaces such as food, vegetation, and clothes. The infectious dose of some pathogens is lower for respiratory tract infections than for infections via the gastrointestinal tract; thus, for some pathogens, inhalation may be a more likely route for disease than either contact or ingestion (Hoadley and Goyal, 1976). Although bacteria and viruses in aerosols have been detected several hundred meters downwind from the point of aerosol generation (Sepp, 1971; Teltsch and Katzenelson, 1978; Johnson et al., 1980; Bausum et al., 1983; Camaan et al., 1986; Camaan et al., 1988), there have not been any documented disease outbreaks in the United States resulting from spray irrigation with treated municipal wastewater that has been disinfected (U.S. Environmental Protection Agency, 1980).

For intermittent spraying using recovered water, occasional inadvertent contact should pose little microbiological health hazard from inhalation. Aerosols emitted from cooling towers used continuously may present a greater concern if the water is not properly disinfected. Although studies indicate that the health risk associated with aerosols from spray irrigation sites is low, the general practice has been to limit exposure to aerosols produced from waters that are likely to contain pathogenic microorganisms until more sensitive and definitive studies can be conducted to fully evaluate the ability of pathogens contained in aerosols to cause disease.

In general, the health hazards associated with the ingestion of inorganic constituents through water or food are well established (U.S. Environmental Protection Agency, 1976), and EPA has set maximum contaminant levels (MCLs) for drinking water. For crop irrigation the elements of greatest concern at elevated levels are cadmium, copper, molybdenum, nickel, and zinc. Nickel and zinc are of a lesser concern than cadmium, copper, and molybdenum because they have visible adverse effects in plants at lower concentrations than the levels harmful to animals and humans. Cadmium, copper, and molybdenum, however, can be harmful to animals at concentrations too low to affect plants.

Copper is not toxic to monogastric animals, but may be toxic to ruminants; however, their tolerance to copper increases as available molybdenum increases (U.S. Environmental Protection Agency, 1981). Molybdenum can also be toxic when available in the absence of copper. Cadmium is of particular concern because it can accumulate in the food chain. It does not adversely affect ruminants in the small amounts they ingest. Most milk and beef products are also unaffected by livestock ingestion of cadmium because it is stored in the liver and kidneys of the animal instead of the fat or muscle tissue.

Pretreatment of waters of impaired quality prior to surface spreading and the additional treatment provided during percolation through the vadose zone are generally sufficient to reduce the concentrations of inorganic constituents of health concern to levels that are acceptable for nonpotable uses of the recovered water. Similarly, the high levels of treatment usually required for injection water reduce inorganic constituents to low levels. Trace elements in treated municipal wastewater normally occur in concentrations of less than a few milligrams per liter, with usual concentrations less than 100 µg/l (Page and Chang, 1985).

Organic constituents may be of concern where recovered water is used for crop irrigation, where the organics may bioaccumulate in the food chain, such as in fish-rearing ponds, or where the water from irrigation or other uses reaches potable supplies, which may result in exposure from direct ingestion or by inhalation or skin contact (Andelman, 1990; Wilkes et al., 1992).

Crop uptake of certain pesticides has been studied (U.S. Environmental Protection Agency, 1973; Palazzo, 1976), and uptake of polychlorinated biphenyls by root crops has been demonstrated under field conditions (Iwata and Gunther, 1976). Uptake of organic compounds is affected by the solubility, size, concentration, and polarity of the organic molecules; the organic content, pH, and microbial activity of the soil; and climate. A study on health risks associated with land application of sludge found that not more than 3 percent or less of the pesticides and herbicides present in the soil passed into plant foliage (Pahren et al., 1979). It has been postulated that most trace organic compounds are too large to pass through the semipermeable membrane of plant roots (U.S. Environmental Protection Agency, 1981).

SUMMARY

To evaluate the health implications of the use of reclaimed water for ground water recharge, information including, but not limited to, the following is needed: types and concentrations of chemicals present in the water (source water, recovered water, water at the point of use); environmental fate, transport, degradation, transformation, and any effect of treatment or processing of the water on the parent compound and breakdown products; toxicological properties of the chemicals; end use of the water; and characterization of human exposure.

There are a number of recharge projects in the United States and other parts of the world which have provided analytical data on the contaminants found in the reclaimed water. The constituents are highly variable, depending on the source of water and specific sites involved. Some of the findings show levels exceeding the U.S. drinking water standards. A more limited number of projects have provided health effects data associated with use of recovered water. The majority of the limited health effects data do not suggest a health concern. Overall continued chemical monitoring and characterization and more toxicological evaluations are needed, however. Exposure data are lacking, and the magnitude of human exposure should be better defined.

Many of the specific conclusions and recommendations made by the Scientific Advisory Panel on Ground Water Recharge (State of California, 1987) regarding toxicological evaluations and epidemiologic studies remain valid. When only mutagenicity data (obtained from the Ames assay) are available, it is clear that the database is not adequate for the assessment of health risks associated with the water. A spectrum of toxicological studies would be needed to provide the necessary data. The design of the studies would depend on what is known (e.g., point of withdrawal, chemical concentration, known constituents) about the water, the selection and preparation of which for testing would be specific for each project. Epidemiological studies can provide a baseline for future references or may assist in identifying a relationship (spatial, temporal, causal) between use of recovered water and associated health effects. However, if not properly designed and conducted with adequate resources, planning, conduct and follow-up, epidemiological studies can be limited by small sample size, existence of confounding factors, and lack of known exposure and control group. They are often of limited value in identifying a cause-and-effect relationship. Therefore, epidemiological studies are not recommended unless the criteria are met for conducting a comprehensive and valid examination.

No quality standards currently exist for chemicals in recovered water intended for potable use. Criteria should be established for the evaluation of the presence of chemicals in such water. Comparison with existing water supply or drinking water standards is one common and convenient approach for evaluating the quality of the recovered water. However, drinking water standards are not based solely on health considerations but also on technical feasibility, detectability, and economic considerations. In addition, there may be other potential sources of exposure to the same chemicals. Therefore, total exposure and methods used to derive health base levels should be considered.

Disinfection by-products (DBPs) are of potential concern in ground water recharge systems used for potable water, as they are in water supplies drawn from surface or non-recharged ground water. The nature and toxicity of such DBPs have been most widely studied for chlorine disinfection of potable water supplies. DBPs from the use of other disinfectants, such as ozone and chloramines, are not as well characterized. Also, the possible differences in the nature and quantities of DBPs from the disinfection of wastewaters that may be used for ground water recharge, such as highly treated sewage and agricultural wastewaters, have for the most part not been studied. In conventional public water supply systems, humic materials are normally the major precursors in the formation of DBPs. In water from unconventional sources, however, there may be a different mix of DBPs, and the nature of the dissolved organic carbon remaining following disinfection may be different from that of humic materials.

As with conventional potable water disinfection, a key issue in developing ground water recharge systems is the need to balance the risks in using chemical disinfectants to reduce the number of pathogenic microorganisms with the risks associated with the DBPs formed in the process. A variety of factors influence the pathogen content of wastewater, including season, the economic status of the population producing the wastewater, water use patterns, and the quality of the potable water supply. Almost half of the outbreaks of waterborne disease in the United States in the 1980s were associated with the use of undisinfected or inadequately disinfected ground water, with sewage most often identified as the source of contamination. The causative agents were identified in less than half of all the outbreaks. The Norwalk virus may have been responsible for many of these outbreaks.

Assessing the risk of an individual from pathogens in recovered water is a difficult and uncertain task. Bacteria and parasites generally are removed to a greater extent than enteric viruses during infiltration through soils; thus viruses are of greater concern when exposure is to the affected ground water. The number and types of pathogens in recovered water depend on the level of treatment the water has received. Studies of enterovirus removal in water reclamation plants in southern California indicate that, with proper treatment, viruses can be reduced to below detectable levels. However, in general there is little information on the concentrations of several pathogens of public health concern in such studies, including hepatitis A and E viruses, rotavirus, Norwalk virus, *Giardia*, and *Cryptosporidium.* Estimates of infection risks from the accidental ingestion of 100 ml of final effluent from advanced municipal wastewater treatment effluent ranged from approximately 1 in 10,000 for *Cryptosporidium* to 2 in 100 million for viruses. Similarly, estimates of annual risks to those exposed to a specific use of ground water recharged with a chlorinated secondary sewage effluent ranged from 8 in 100 million to 1.5 in billion. Currently, the control of such risks below 1 in 10,000 is considered acceptable.

On the basis of available information, there is no indication that the health risks from using reclaimed wastewater are greater than those from using existing water supplies or that the concentrations of chemicals, with several exceptions, or microorganisms are higher than those established in drinking water standards set by EPA. One limitation is that not all the chemicals identified have drinking water standards for such a comparison. There are other uncertainties, too, in such an evaluation, including the limited chemical and toxicological characterizations of source and recovered waters, and the uncertain environmental fates of the chemicals and microorganisms in the recharge systems, and the limited epidemiological data. Furthermore, it should be remembered that the EPA standards are based on water sampled from high quality sources. Such standards are often years behind our knowledge and our knowledge at any time is limited. (For example, health effects are determined for each organic compound separately, not for the inevitable mixtures of organics.) Accordingly, monitoring of potentially toxic constituents and pathogenic microorganisms should be required in using water extracted from these recharge systems. Increased reliance on technology also brings increased uncertainty and thus some risk. Given these uncertainties, where recovered recharge water is intended for potable purposes and human exposure is increased, great care is necessary (Asano et al., 1992; NRC, 1983; NRC, 1994).

REFERENCES

Amy, G. L., J. M. Thompson, M. K. Davis, and S. W. Krasner. 1990. Evaluation of THM preprecursor contributions from agricultural drains. J. Am. Water Works Assoc. 82(1):57-64.

Ayers, R. S., and D. W. Westcott. 1985. Water Quality for Agriculture. FAO Irrigation and Drainage Paper 29, Rev. 1. United Nations Food and Agriculture. Rome.

Asano, T., L. Y. C. Leong, M. G. Rigby, and R. H. Sakaji. 1992. Evaluation of the California wastewater reclamation criteria using enteric virus monitoring data. Water Sci. Technol. 26:1513-1524.

Bausum, H. T., S. A. Schaub, R. E. Bates, H. L. McKim, P. W. Schumacher, and B. E. Brockett. 1983. Microbiological Aerosols From a Field-Source Wastewater Irrigation System. J. Water Pollution Control Federal, 55(1):65-75.

Bryan, F. L. 1974. Diseases Transmitted by Foods Contaminated by Wastewater. In: Wastewater Use in the Production of Food and Fiber. EPA-660/2-74-041. U.S. Environmental Protection Agency. Washington, D.C.

Bull R. J., C. Gerba, and R. R. Trussell. 1990. Evaluation of the health risks associated with disinfection. Critical Reviews in Environmental Control 20:77-113.

Bull R. J., and F. C. Kopfler. 1991. Health Effects of Disinfectants and Disinfection Byproducts. Denver, Colo.: American Water Works Association Research Foundation.

Camaan, D. E., R. J. Graham, M. N. Guentzel, H. J. Harding, K. T. Kimball, B. E. Moore, R. L.

Northrop, N. L. Altman, R. B. Harrist, A. H. Holguin, R. L. Mason, C. B. Popescu, and C. A. Sorber. 1986. The Lubbock Land Treatment System Research and Demonstration Project: Volume IV. Lubbock Infection Surveillance Study. EPA-600/2-86-027d. Health Effects Research Laboratory, U. S. Environmental Protection Agency. Research Triangle Park, N.C.

Camaan, D. E., B. E. Moore, H. J. Harding, and C. A. Sorber. 1988. Microorganism Levels in Air Near Spray Irrigation of Municipal Wastewater: the Lubbock Infection Surveillance Study. J. Water Pollution Control Federal 60(11):1960-1970.

Craun, G. F. 1990. Methods for investigation and prevention of waterborne disease outbreaks. Report No. EPA-600/1-90/005a. U.S. Environmental Protection Agency, Office of Research and Development. Cincinnati, Ohio.

Craun, G. F. 1991. Causes of waterborne outbreaks in the United States. Water Sci. Technol. 24:17-20.

Feachem, R. G., D. J. Bradley, H. Garelick, and D. D. Mara. 1983. Sanitation and Disease-Health Aspects of Excreta and Wastewater Management. Chichester, England: John Wiley, The World Bank.

Florida Department of Environmental Regulation. 1990. Reuse of Reclaimed Water and Land Application. Chapter 17-610, Florida Administrative Code. Tallahassee, Fla.

Gerba, C. P., and S. M. Goyal. 1985. Pathogen removal from wastewater during groundwater recharge. Pp. 283-317 in Artificial Recharge of Groundwater, T. Asano, ed. Boston, Mass: Butterworth.

Gerba, C. P., and J. B. Rose. 1993. Estimating viral disease risk from drinking water. Pp.117-135 in Comparative Environmental Risk Assessment, C. R. Cothern, ed. Boca Raton, Fla: Lewis Publishers.

Haas, C. N. 1983. Wastewater disinfection and infectious disease risks. Crit. Rev. Environ. Contr. 17:1-20.

Herwaldt, B. L., G. F. Craun, S. L. Stokes, and D. D. Juranek. 1992. Outbreaks of waterborne disease in the United States: 1989-1990. J. Am. Water Works Assoc. 84:129-135.

Herren-Freund, S. L., M. A. Pereira, and G. Olson. 1987. The carcinogenicity of trichloroethylene and its metabolites, trichloroacetic acid and dichloroacetic acid, in mouse liver. Toxicol. Appl. Pharmacol. 90:83.

Hoadley, A. W., and S. M. Goyal. 1976. Public health implications of the application of wastewater to land. P. 1092 in Land Treatment and Disposal of Municipal and Industrial Wastewater, R. L. Sanks and T. Asano, eds. Ann Arbor, Mich.: Ann Arbor Science Publishers.

International Agency for Research on Cancer. 1987. Overall evaluations of carcinogenicity: An updating of IARC Monographs Volumes 1 to 42. IARC Monographs on the Evaluation of the Carcinogenic Risk of Chemicals to Man. Suppl. 7. Lyons, France: International Agency for Research on Cancer.

International Agency for Research on Cancer. 1991. IARC Monographs on the Evaluationof Carcinogenic Risks to Humans. Chlorinated Drinking-water; Chlorination By-products; Some Other Halogenated Compounds; Cobalt and Cobalt Compounds, Vol. 52. Lyons, France: International Agency for Research on Cancer, World Health Organization.

Iwata, Y., and F. A. Gunther. 1976. Translocation of the polychlorinated biphenyl oroclor 1254 from soil into carrots under field conditions. Arch. Environ. Contam. Toxicol. 4(1):44-59.

Johnson, D. E., D. E. Camaan, J. W. Register, R. E. Thomas, C. A. Sorber, M. N. Guentzel, J. M. Taylor, and W. J. Harding. 1980. The Evaluation of Microbiological Aerosols Associated with the Application of Wastewater to Land: Pleasanton, CA. EPA-600/1-80-015. U.S. Environmental Protection Agency. Cincinnati, Ohio.

Kaplan, J. E., G. W. Gary, R. C. Baron, W. Singh, L. B. Schonberger, R. Feldman, and H. Greenberg. 1982. Epidemiology of Norwalk gastroenteritis and the role of Norwalk virus in outbreaks of acute nonbacterial gastroenteritis. Ann. Intern. Med. 96:756-761.

Keswick, B. H., T. K. Satterwhite, P. C. Johnson, H. L. DuPont, S. L. Secor, J. A. Bitsura, G. W.

Gary, and J. C. Hoff. 1985. Inactivation of Norwalk virus in drinking water by chlorine. Appl. Environ. Microbiol. 50:261-264.

Kronberg L., R. F. Christman, R. Singh, and L. M. Ball. 1991. Identification of oxidized and reduced forms of the strong bacterial mutagen(z)-2-chloro-3-(dichloromethyl)-4-oxobutenoic acid (MX) in extracts of chlorine treated water. Environ. Sci. Technol. 25:99-104.

Lauer, W. C., F. J. Johns, G. W. Wolfe, B. A. Myers, L. W. Condie, and J. F. Borzelleca.1990. Comprehensive Health Effects Testing Program for Denver's Potable Water Reuse Demonstration Project. J. Toxicol. Environ. Health 30:305-321.

Murphy, W. H., and J. T. Syverton. 1958. Absorption and translocation of mammalian viruses by plants. II. Recovery and distribution of viruses in plants. Virology 6(3):623.

National Research Council. 1983. Risk Assessment in the Federal Government: Managing the Process. Washington, D.C.: National Academy Press.

National Research Council. 1994. Science and Judgment in Risk Assessment. Committee on Risk Assessment of Hazardous Air Pollutants. Washington, D.C.: National Academy Press.

Nellor, M. H., R. B. Baird, and J. R. Smyth. 1984. Summary. Health Effects Study—Final Report. County Sanitation Districts of Los Angeles County. Whittier, Calif.

Office of Environmental Health Hazard Assessment. 1992. Recommended Public Health Level for Arsenic in Drinking Water. California Environmental Protection Agency, Office of Environmental Health Hazard Assessment. Berkeley, Calif.

Page, A. L., and A. C. Chang. 1985. Fate of wastewater constituents in soil and groundwater: trace elements. Pp. 13-16 in Irrigation with Reclaimed Municipal Wastewater—A Guidance Manual. G. S Pettygrove and T. Asano, eds. Prepared by the California State Water Resources Control Board. Chelsea, Mich.: Lewis Publishers.

Pahren, H. R., J. B. Lucas, J. A. Ryan, and G. K. Dotson. 1979. Health risks associated with land application of municipal sludge. J. Water Pollution Control Federation, 51(11):2588-2601.

Palazzo, A. J. 1976. The Effects of Wastewater Applications on the Growth and Chemical Composition of Forages. Report 76-9. Cold Regions Research and Engineering Laboratory, U.S. Army Corps of Engineers. Hanover, NH.

Parnell, M. J., L. D. Koller, J. H. Exon, and J. M. Arnzen. 1986. Trichloroacetic acid effects on rat liver peroxisomes and enzyme-altered foci. Environ. Health perspect. 69:73-79.

Pawlowski, S. 1992. Proposed revisions to Arizona's water reuse regulations. Paper presented at the Salt River Project Water Reuse Symposium, Nov. 2, 1992, Tempe, Ariz.

Payment, P., L. Richardson, M. Edwardes, E. Franco, and J. Siemiatycki. 1991. A prospective epidemiological study of drinking water related gastrointestinal illness. Water Sci. Technol. 24:27-28.

Pickard, D., D. Bracciano, and D. Holmes. Undated. Determination of Anticipated Health Effects for Reuse of Municipal Wastewater. Department of Sanitary Services, Tampa, Fla.

Putnam, S. W., and J. D. Graham. 1993. Chemicals versus microbials in drinking water: A decision sciences perspective. J. Am. Water. Works Assoc. 85:57-61.

Rao, V. C., and J. L. Melnick. 1986. Environmental Virology. American Society for Microbiology. Washington, D.C.

Regli, S., J. B. Rose, C. N. Haas, and C. P. Gerba. 1991. Modeling the risk from Giardia and viruses in drinking water. J. Am. Water Works Assoc. 83:76-84.

Rose, J. B., and C. P. Gerba. 1991. Use of risk assessment for development of microbial standards. Water Sci. and Technol. 24(2):29-34.

Rose, J. B., and R. P. Carnahan. 1992. Pathogen removal by full scale wastewater treatment. Prepared for Florida Department of Environmental Regulation. Tallahassee, Fla.

Sepp, E. 1971. The Use of Sewage for Irrigation—A Literature Review. Bureau of Sanitary Engineering, California Department of Public Health. Berkeley, Calif.

Shuval, H. I. 1978. Land treatment of wastewater in Israel. Pp. 429-436 in State of Knowledge in Land Treatment of Wastewater, Vol. 1. Proceedings of an International Symposium. Cold Regions Research and Engineering Laboratory, U.S. Army Corps of Engineers. Hanover, NH.

Sobsey, M. D. 1989. Inactivation of health-related microorganisms in water by disinfection processes. Water Sci. Technol. 21:179-195.

Sobsey, M. D., T. Fuji, and R. M. Hall. 1991. Inactivation of cell-associated and dispersed hepatitis A virus in water. J. Am. Water Works Assoc. 83:64-67.

State of California. 1978. Wastewater Reclamation Criteria. California Administrative Code, Title 22, Division 4. California Department of Health Services, Sanitary Engineering Section, Berkeley, California.

State of California. 1987. Report of the Scientific Advisory Panel on Groundwater Recharge with Reclaimed Wastewater. Department of Water Resources and Department of Health Services, State Water Resources Control Board. Sacramento, Calif.

State of Texas. 1990. Use of Reclaimed Water. Texas Administrative Code, Chapter 310, Subchapter A. Texas Water Commission. Austin, Tex.

State of Washington. 1993. Water Reclamation and Reuse Interim Standards. Department of Health, State of Washington. Spokane, Wash.

Straub, T. M., I. L. Papper, and C. P. Gerba. 1993. Hazards from pathogenic microorganisms in land-disposed sewage sludge. Rev. of Environ. Contam. in Toxicol. 132:55-91.

Tanaka, H., T. Asano, E. D. Schroeder, and G. Tchobanoglous. 1993. Estimating the reliability of wastewater reclamation and reuse using enteric virus monitoring data. Paper presented at the 66th Water Environment Federation Annual Conference and Exposition, Oct. 3-7, Anaheim, Calif. 14 pp.

Teltsch, B., and E. Katzenelson. 1978. Airborne Enteric Bacteria and Viruses from Spray Irrigation with Wastewater. Appl. Environ. Microbiol. 35(2):290-296.

U.S. Environmental Protection Agency. 1973. Water Quality Criteria 1972. EPA-R3-73-033. A report of the Committee on Water Quality Criteria, National Academy of Sciences—National Academy of Engineering. Washington, D.C.

U.S. Environmental Protection Agency. 1976. Quality Criteria for Water. U.S. Environmental Protection Agency. Washington, D.C.

U.S. Environmental Protection Agency. 1980. Wastewater Aerosols and Disease. EPA-600/9-80-028. Proceedings of Symposium, Sept. 19-21, 1979. H. Pahren and W. Jakubowski, eds. Health Effects Research Laboratory, U.S. Environmental Protection Agency. Cincinnati, Ohio.

U.S. Environmental Protection Agency. 1981. Process Design Manual: Land Treatment of Municipal Wastewater. EPA-625/1-81-013. Center for Environmental Research Information, U.S. Environmental Protection Agency. Cincinnati, Ohio.

U.S. Environmental Protection Agency. 1986. Drinking Water Criteria Document for Chlorophenols. External review draft. ECAO-CIN-D005. U.S. Environmental Protection Agency. Washington, D.C.

U.S. Environmental Protection Agency. 1992. Guidelines for Water Reuse. EPA/625/R-92/004. Center for Environmental Resources Information, Office of Technology Transfer and Regulatory Support. Cinncinnati, Ohio.

U.S. Environmental Protection Agency. 1993a. Integrated Risk Information system (IRIS). U.S. Environmental Protection Agency. Washington D.C.

U.S. Environmental Protection Agency. 1993b. Drinking Water Regulations and Health Advisories. Office of Water, U.S. Environmental Protection Agency. Washington, D.C.

Western Consortium for Public Health. 1992. City of San Diego Total Resource Recovery Project. Health Effects Study. Western Consortium for Public Health. Berkeley, Calif.

Wilkes, C. R., M. J. Small, J. B. Andelman, N. J. Giardino, and J. Marshall. 1992. Inhalation exposure model for volatile chemicals from indoor water uses. Atmos. Environ. 26A:2227-2236.

Yanko, W. A. 1993. Analysis of 10 years of virus monitoring data from Los Angeles County treatment plants meeting California wastewater reclamation criteria. Water Environ. Res. 65:221-226.

5

Economic, Legal, and Institutional Considerations

Decision making by public and private entities about water resources cannot be understood without consideration of the relevant institutional factors. Indeed, many water resource professionals would agree that "institutional problems in water resources development and management are more prominent, persistent, and perplexing than technical, physical, or even economic problems..." (Ingram et al., 1984). Because ground water recharge projects are still somewhat novel, these factors are less obvious, than, for example, the familiar dynamics that led to western dams and irrigation projects. This chapter examines the economics, regulatory schemes, and key actors that affect ground water recharge projects.

ECONOMIC ISSUES

The scarcity of high-quality water supplies is intensifying in many regions of the United States. Environmental and fiscal problems constrain the construction of new surface water storage facilities. In many regions, declining water quality threatens potable supplies as well as potential sources of new supply. The constraints on the development of additional supplies have become more stringent at the same time that demands for additional municipal and industrial water are growing.

Simultaneously, water quality laws, policies, and regulations have forced municipalities to subject wastewater to increasingly more expensive treatment processes prior to discharge to surface waters. Because of this required treatment, the quality of many treated municipal wastewaters is sufficiently high that with relatively modest additional treatment they can be recycled and made available for a variety of uses. The economic attractiveness of treated municipal wastewater as a source of supply has focused increasing attention on the possibility of using it to recharge aquifers. Historically, treated wastewater has been used on a modest scale both to augment ground water supplies and as a means of protecting aquifers in coastal regions from seawater intrusion.

The Economics of Ground Water Use

There is a substantial and varied literature on the economics of ground water use (see, for example, Burt, 1970; Cummings, 1970; Gisser, 1983; Burness and Martin, 1988; Provencher and Burt, 1993). A number of common principles related to the use and management of ground water are developed and characterized in this literature. For example, ground water is most efficiently used when it is extracted at rates such that the net benefits (total benefits net of total costs) from use are maximized over time. The benefits are typically determined by the use to which the water is put. In the short term, costs include the cost of extracting ground water and the opportunity, or user, cost.

The cost of extracting ground water is usually a function of energy cost, pump efficiency, and the depth from which the water must be pumped. Extraction cost increases as energy cost and pumping depth increase, and it declines as pump efficiency increases. The opportunity cost of extraction is the cost of extracting the water now rather than leaving it for later use. The opportunity cost, which is frequently called a user cost, captures the fact that water pumped in the current period results in a lowered ground water table for all future periods if pumping rates exceed safe yields of the aquifer. The incremental cost of pumping from a lowered water table in the future must be accounted for if current extractions are to be economically efficient. Much of the economic literature on ground water resources focuses on the fact that where ground water is treated as a common property resource, extractions tend to occur at rates that are inefficient. When pumpers fail to account for all of the costs of extraction, including the user cost, the rates of extraction are greater than the economically efficient rate.

In the long run, the rates of extraction for any given aquifer cannot exceed the rate at which the aquifer is recharged—the safe yield—without overdrafting the aquifer. Overdrafting can brings costs: land subsidence, greater risk of flooding, greater risk of salt water intrusion in coastal areas, and the increased costs of reaching and pumping the water from the lowered water table. When overdrafting is persistent, the ground water table is progressively lowered until a point is reached at which the cost of extracting the ground water from any lower depth is greater than the benefits that could be obtained from any of the uses to which that water might be put. At this point, it is no longer economical to continue pumping and further declines in the ground water table are arrested. Ultimately, proper management of the relative magnitudes of the pumping cost and the benefits from use can ensure that only the annual recharge is extracted; when this balance is reached, the aquifer is said to be in steady state.

There can be circumstances, however, in which overdrafting is economically efficient. For instance, when the benefits of use are quite high in relation to the costs of extraction, overdrafting may be justified. It is important to remember, however, that overdrafting ultimately is self-terminating.

Another consideration relates to the optimal water table depth at which steady state is reached. The optimal steady-state depth will be attained if all costs of extraction, including the user cost, are accounted for by pumpers. Where ground water is treated as a common property resource, pumpers have an incentive to ignore the user cost, and this results in a deeper than optimal steady state depth. In this instance, rates of extraction leading to the steady-state are greater than the economically optimal rate, and pumping depths are lower overall than the optimal pumping depth.

The principal lesson from the basic economics of ground water use is that where ground water is exploited in an individualistically competitive fashion, the rates of extraction and the steady state depth tend not to be economically optimal. The rule of capture prevails, and this means that pumpers only obtain the right to use ground water once they have pumped it and are prepared to put it to use. The user cost tends to be ignored, both because pumpers believe that their own extractions will have an infinitesimally small impact on other pumpers and because they perceive that voluntary restraints on extractions serve only to make the water available to potentially competing pumpers. Corrective measures that have been identified include pumping quotas, pump taxes equivalent to the marginal user cost, and the formal vesting of property rights to ground water in situ.

Understanding of the economics of ground water use leads to the conclusion that ground water will not be used in an economically efficient fashion when it is treated as a common property resource. A single pumper will usually account for the consequences of today's extractions on tomorrow's extraction costs. The problem will arise where there are many pumpers behaving competitively. To address it, some form of management will be required.

This lesson has important implications for the artificial recharge of ground water. If ground water is treated as a common property resource, the incentive to incur the expenses associated with artificial recharge will be eroded because the additional water will be available for capture by other pumpers, who presumably are not obligated to help pay for the recharge operation. Thus, in situations where there are many competing pumpers and no regulation of extractions, the returns from artificial recharge operations cannot be fully captured by those who plan and finance the operations.

Artificial recharge of ground water can have at least two distinct purposes. First, the recharge water can be extracted and put to direct use. Second, in situations where an aquifer may be threatened by seawater intrusion or intrusion of very low quality ground water, artificial recharge can be used to protect the entire aquifer. The economics of artificial recharge for direct use has been analyzed in some detail by Brown and Deacon (1972), Cummings (1971), and Vaux (1985). Artificial recharge augments the rate of recharge and thus increases quantities of water that can be optimally extracted at any point in time. Moreover, artificial recharge results in an optimal steady state pumping depth that is shallower than the optimal depth that would have prevailed in the absence of artificial recharge. Finally, artificial recharge can ameliorate problems of uncertainty that arise from the inherent variability of runoff and surface supplies that contribute to or are available for ground water recharge. These conclusions require, of course, that the incremental costs of the artificial recharge water be less than or equal to the incremental benefits that accrue from the uses to which the water is ultimately put.

The economics of recharge in the second case, where recharge is used to prevent saltwater intrusion or otherwise protect an aquifer, has been treated in a number of case studies (Cummings, 1971; Warren et al., 1975). In these instances, the benefits that accrue are equivalent to the net benefits from the protected aquifer that would be lost if saltwater intrusion remained unabated. An economically efficient recharge operation requires that the incremental costs of recharge be less than or equal to the net benefits protected. If the aquifer to be protected is exploited competitively without pumping restrictions, the net benefits of protection would be less than they would be if extractions were economically optimal.

The Economics of Artificial Ground Water Recharge with Treated Municipal Wastewater

The economic feasibility of ground water recharge with treated municipal wastewater will vary from situation to situation. Recharge is but one option for managing water supply and the disposal of wastewater. The economic feasibility of recharge with treated wastewater will depend critically on the costs of other options for augmenting water supplies, the costs of alternative means for disposing of wastewater, and the benefits or returns that accrue from the availability of additional water supplies and the effective management of wastewater. Thus, in every situation economic feasibility must be assessed within the context of the particular water supply and demand situation and with specific reference to the array of alternatives that may be available to solve the water management problems in question.

Demand

The benefits of additional water supplies are normally measured by the willingness of consumers to pay or by the demand for additional water supplies.

Generally, the willingness to pay for additional municipal and industrial supplies is expected to grow as urban population and economic activity increase. This is particularly true in the arid and semiarid western states, where nearly all of the economic and population growth is occurring in urban areas, but it is also clearly apparent in other water-short areas such as Florida. If the willingness to pay for recharged wastewater exceeds the cost of supplying that water and there are no cheaper alternative sources, recharge with wastewater may be an attractive option.

Even where water quality and other factors are roughly comparable between ground water and alternative sources of surface water, there is at least one reason why the willingness to pay to acquire rights to ground water resources may be higher than for rights to surface water. In many areas, high quality ground water may be economically more attractive than alternative sources of surface water because it is reliably available. In the short run, the availability of ground water is not normally dependent on precipitation in the same way the surface water availability is. Thus, ground water tends to be relatively insulated from the effects of drought and, other things being equal, the willingness to pay for reliable ground water may be higher than for a source subject to interruption. The willingness to pay for reliable ground water sources can be diminished, however, if the quality of the ground water in question is distinctly lower than the quality of comparable surface water supplies or if the risks and uncertainties of adverse effects on human or environmental health associated with recharged ground water supplies are significantly higher than those associated with comparable surface water supplies.

The Cost of Water Supplies

The attractiveness of treated wastewater as a source of ground water recharge depends crucially on how the cost compares to the cost of alternative sources of supply. There are several reasons for believing that treated wastewaters may enjoy considerable cost advantages over other sources in the immediate future.

In most regions of the country, the cost of developing new surface supplies has become prohibitive. Over the last several decades, the cost of constructing civil works has risen faster than the rate of inflation. Moreover, nearly all of the easily developed surface water storage sites have already been developed, leaving only sites that are costlier to develop or quite remote. In many instances, the combination of these two factors alone means that the cost of new surface water storage facilities outstrips the willingness to pay for new supplies. In addition, the willingness to subsidize the cost of new water supply facilities from public revenues has declined dramatically from the dam-building heyday of the 1950s and 1960s and before.

To these financial constraints, the environmental cost of surface water development must be added. It is now recognized that surface water impoundment facilities can cause significant environmental damage. Strong public preferences for environmental amenities, together with the substantial cost of mitigating or compensating for environmental damage, have helped to make the construction of new surface water storage facilities far less attractive than it once was. Concerns about adverse environmental impacts lead to strong political opposition to the development of additional impoundment facilities. Even in the relatively rare instances in which the cost of new facilities is consistent with the willingness to pay, the potential adverse environmental impacts create strong political resistance to the development of such new facilities.

The result is that new surface water storage and conveyance facilities have become a relatively unattractive and, in some instances, unacceptable means of developing new water supplies. On the other hand, many aquifers contain unused storage capacity, which can be developed at relatively modest cost and without the adverse environmental consequences frequently associated with surface storage. Because ground water storage avoids many of the high costs associated with surface storage, conjunctive use—the integrated management of ground and surface water—has become an increasingly attractive option for augmenting developed water supplies. Conjunctive use schemes often prove infeasible, however, because of the absence of "surplus" surface water that could be stored in the ground.

Many western streams are already fully allocated, and the increasing contentiousness of reallocation proposals in a number of middle western and eastern streams suggests that streams in those regions may be fully subscribed *de facto*. The resultant scarcity of surface waters available for development means that the search for additional water supplies has turned to more exotic sources, including the desalinization of brackish waters and seawater. It is true that changes in water allocation institutions, including, for example, more widespread adoption of water markets, may help ameliorate the scarcity of water, particularly in the West. Nevertheless, where new supplies are sought, the costs of treated wastewater may compare favorably with the cost of water from alternative sources because state and federal regulations require extensive treatment of wastewater irrespective of whether and how it is to be reused. The incremental costs of rendering treated wastewater suitable for nonpotable and even potable uses may thus be quite modest in many cases.

The incremental costs of upgrading the quality of treated wastewater will vary widely from situation to situation and depend on a number of variables. The cost of land acquisition for the siting of treatment plants and spreading grounds and the cost of constructing injection wells will vary from project to project and can be significant. Similarly, the need to transport treated wastewater from the place of treatment to the site where it is to be spread or injected may add substantial cost (Vaux, 1985). In many instances, however, the most important determinant of economic feasibility will be the cost of the additional treatment required to improve the quality of the wastewater to the desired level.

Data recently developed by the Orange County Water District (Orange County Water District/County Sanitation Districts of Orange County, 1993) for three different treatment options are illustrative. Table 5.1 presents the quality parameters of the product waters associated with the three treatment options and the costs of those options. The three options entail differences only in the levels of reverse osmosis and microfiltration applied to the product water. For options 1 and 2, it is assumed that only partial microfiltration is provided, with option 2 requiring it to a greater degree than option 1. Option 3 entails full microfiltration. The major quality difference is in the total organic carbon (TOC) reduction.

The data illustrate the sensitivity of wastewater reclamation cost to the level of quality desired. The cost is, of course, also quite sensitive to the difference between the quality of the source water and the desired quality of the product water. Thus, the magnitude of the cost will be crucially conditioned by regulatory requirements on the quality of waters to be reclaimed by spreading or direct injection.

In the Orange County situation the least-cost alternative source of water is surface water delivered from the Metropolitan Water District at a cost of $600 per acre-foot. To this must be added the capital and operating costs of transport-

TABLE 5.1 Comparison of Treatment Costs and Product Water Quality, Well Injection at Orange County, California

Chemical Constituent	Option 1	Option 2[a]	Option 3[a]
Total dissolved solids[b]	650 mg/l	600 mg/l	600 mg/l
Sodium	143 mg/l	139 mg/l	139 mg/l
Chloride	151 mg/l	140 mg/l	140 mg/l
Sulfate	140 mg/l	122 mg/l	122 mg/l
Total organic carbon	8.6 mg/l	5.5 mg/l	5.5 mg/l
Cost per acre-foot[c]	$251	$359	$387

[a]The cost difference between options 2 and 3 is attributable to employment of full microfiltration for option 3 but only partial microfiltration for option 2.

[b]Source water is assumed to contain 900 mg/l total dissolved solids for all three options.

[c]Cost includes capital consumption, debt service, and operation and maintenance costs.

Source: Orange County Water District/County Sanitation Districts of Orange County, 1993.

ing the water to spreading grounds, which are $82.40 per acre foot. The total costs of the least cost alternative, then, is $682.40. Thus, option 1 is the least-cost alternative, but both options 2 and 3 are less costly than purchasing supplemental surface water supplies (Orange County Water District/County Sanitation Districts of Orange County, 1993).

It is important to note that wastewater treatment facilities generally exhibit increasing returns to scale. Thus the costs of operation are usually cited for some constant and optimal quantity of water to be treated, given the size of the treatment facility in question. If the volumes of wastewater to be treated are highly variable from season-to-season or year-to-year, the costs of treatment may be higher since the facilities cannot be operated consistently at optimal levels. Similarly, if the volumes of wastewater to be treated decline over time because of water conservation efforts, the costs of wastewater treatment could increase as the volumes of water to be treated no longer match well with the size or scale of the treatment facility. These points underscore the importance of accounting for likely changes in future volumes of wastewater and the constancy of those volumes in the design of wastewater treatment facilities.

Inasmuch as the costs and benefits of using reclaimed wastewater vary from situation to situation, it is difficult to generalize meaningfully about the economic attractiveness of reclaimed wastewater. It is clear, however, that the feasibility of artificial recharge requires the existence of unambiguous rights to the reclaimed water once it is in the ground. Beyond this, the specific economic calculus depends on the cost of the particular treatment and spreading or reinjection scheme as well as the cost of alternative sources of supply.

LEGAL ISSUES

Artificial recharge of ground water is one of the developments in water management that is challenging existing legal strictures to respond to changing societal needs. Experience with recharge projects suggests that society's laws can evolve to accommodate new strategies such as artificial recharge, if demand is strong enough.

One difficult question raised by ground water recharge is, What policies should be formulated to protect public health, safety, property, third-party, and ecological interests, while not imposing inappropriate controls on this form of water development? The decentralized nature of the current regulatory structure will, in time, provide experience that demonstrates the merits of different regulatory standards. For now, some of the regulations that have been applied by different government entities can be surveyed and guidance sought from these policies. The controls imposed vary greatly from jurisdiction to jurisdiction, so what follows is a conceptual guide to the sort of considerations that have been raised in different areas. California has the most extensive regulatory scheme for ground water recharge, and these regulations and their institutional setting are illustrative.

This study is focused on the artificial recharge of ground water for subsequent reuse of water, whether for potable or nonpotable uses. The legal issues are grouped under the general topics of water rights, protection of ground water quality, use of the recharge water after recovery, and environmental consequences. General statutes that might have a bearing on projects are also reviewed. Considerations that are unique to the source of the recharge water (e.g., treated municipal wastewater, stormwater runoff, and irrigation return flow) are addressed when possible.

Water Rights

A prime issue in ground water recharge is the ownership of the water proposed for recharge. A project proponent must have the legal right to use the source water for that purpose. As a corollary, the project proponent must have the legal right, against other competing users, to withdraw the recharge water.

A water right is commonly established for some use, such as irrigation, industrial processing, or domestic water supply. When the source of ground water recharge is water that has previously been used in some fashion, the question presented is whether the entitlement to use it also creates a right to control what is left over. For example, domestic wastewater can be viewed as a liability that a city must dispose of, or as an asset that might be of value to someone. In the arid West, the only flow in a stream may come from wastewater produced and discharged by a city. Downstream users may become dependent on this flow, and ecosystems may emerge that rely on it. Someone proposing to use this "resource" for a new purpose, such as ground water recharge, must have a legal entitlement to use the water in that way.

The ownership of wastewater was at issue in *Arizona Public Service Co. v. Long*, 160 Ariz. 429, 773 P.2d 988 (1989). In this case, a municipal government was challenged on its ability to sell effluent that it had formerly disposed of in a stream. The court held that the effluent could be sold by the city and that the city was not required to continue the discharge. The water was neither "surface water" nor "ground water," although the legislature could in the future bring it within its statutory scheme for these waters (McGinnis, 1990). The New Mexico Supreme Court in *Reynolds v. City of Roswell*, 99 N.M. 84, 654 P.2d 537 (1982), also permitted a discharger to cease discharging effluent to a stream, despite the objections of the state engineer. New Mexico, in fact, has a statute defining effluent as "private waters" and allowing the discharger to reuse it (N.M. Stat. Ann. § 72-5-27).

The California legislature established a statutory right to reclaimed wastewater in the entity that operates a wastewater treatment facility (Cal. Water Code § 1210). The statute establishes rights in the wastewater with respect to the supplier of the water. With respect to downstream entities that have legal interests in flows, however, permission of the water board must be obtained before the discharge is affected (Cal. Water Code § 1211).

The existence of a right to reuse water may also be dependent on the nature of the original entitlement to water. A user's water right may have been calculated assuming that a certain volume of water would be returned to the stream, so that the user would not have the right to instead divert that water. In some instances, a user's entitlement to reuse water will be a matter of contractual arrangements with the supplier.

Finally, in some states an additional element of ownership may be required before a project can proceed. Under the prior appropriation doctrine, a use must be permissible, or "beneficial," under state law for an applicant to have rights to the water. Colorado, Idaho, Kansas, Nebraska, Oklahoma, and Oregon have recognized ground water recharge in their water laws in varying manners (Colo. Rev. Stat. § 37-92-103(10.5) (1992); Idaho Code § 42-4201A(a)(2) (1992); Kan. Stat. Ann. § 82a-928 (1992); Neb. Rev. Stat. § 46-295 (1992); Okla. Stat. Title 82, §§ 1020.1-1020.22 (1992); Or. Rev. Stat. § 537.135 (1991)). Florida takes an inventive approach to linking water rights to reuse of reclaimed wastewater, including reuse through recharge of ground water. Regulations of the Environmental Regulation Commission provide that "In implementing consumptive use permitting, a reasonable amount of reuse of reclaimed water from domestic wastewater treatment facilities shall be required within designated water resource caution areas, considering economic, environmental, and technical factors" (Florida Administrative Code A.R. 17-40.416(2)).

Environmental statutes also may affect a user's right to cease making a discharge to a stream or wetland for the purpose of ground water recharge. A new diversion could, for example, affect an endangered species (Western States Water Council, 1990), but aside from protections offered endangered species there would generally be no protection for the affected environment. Oregon is an exception to this general rule: it has provided by statute that a ground water recharge permit shall not be issued "unless the supplying stream has a minimum perennial stream flow established for the protection of aquatic and fish life" (Or. Rev. Stat. § 537.135 (1991)). An environmental review, if required under state or federal law, might identify the environmental interest in maintaining a discharge. Interestingly, California has acted to protect two environmental opportunities that can be created by water reuse: if water reuse lessens stream demand, the remaining flow in the stream can be protected from new appropriations, and if reclaimed water is discharged to a stream, the newly created streamflow can be protected (Cal. Water Code § 1212).

Water rights are also of key importance when the time comes to withdraw water from a ground water basin. Without clear protection of the project proponent's economic investment in the recharge water, there is no incentive for projects. A study by the Western States Water Council identified only a few western states that explicitly protect rights in recharge water (Western States Water Council, 1990). In California, case law has established the nature of the recharger's right to exclusive withdrawal of the recharged ground water (*Los Angeles v. San Fernando*, 14 Cal. 3d 199, 537 P.2d 1250 (1975) and *Alameda County Water District v. Niles Sand and Gravel Co.*, 37 Cal. App.3d 924, 112 Cal. Rptr. 846 (1974), cert. denied, 419 U.S. 869 (1975)).

Protection of Ground Water Quality

The addition of water to ground water can affect the quality of the "native" ground water. Ground water quality is the focus of much of environmental law, but there is no comprehensive federal ground water statute.

Congress has, however, given EPA the ability to regulate certain types of ground water recharge through the Underground Injection Control (UIC) program of the Safe Drinking Water Act (42 U.S.C. §§ 300h to 300h-7 (1988)). This act is administered by EPA and by states with approved programs. The act does not protect all ground water, but rather protects, as its name suggests, underground sources of drinking water (USDW), which are aquifers that are used or could be used for public water systems. While there is a general presumption that aquifers of good-quality water are USDWs, the regulations allow exemptions primarily where the aquifer is of poor quality (40 CFR § 146.3 (1992); 40 CFR § 146.4 (1992)).

This authority extends to two significant forms of ground water recharge: injection wells for highly treated wastewater and dry wells used to dispose of stormwater runoff. Both are Class V wells, within the terminology of the act. Dry wells are brought within the scheme by the inclusion of "[a]ny dug hole or well that is deeper than its largest surface dimension, where the principal function of the hole is emplacement of fluids" (40 CFR § 144.1(g)(1) (1992)). The dispositive issue controlling whether remediated wastewater is brought within the regulatory system is thus the diameter of the well. Residential septic systems are specifically excluded from the regulations (40 CFR § 144.1(g)(2) (1992)).

The regulatory authority given to EPA by the act for Class V wells has not been exercised by the agency, and the regulations now require little more than notification of the entity administering the program and submission of certain information (40 CFR § 146.52(a) (1992); 40 CFR § 144.24 (1992)). The agency administering the program is given authority to take action when a Class V well "may cause a violation of primary drinking water regulations" or where it "may be otherwise adversely affecting the health of persons" (40 CFR § 144.12 (1992)). EPA has yet to exercise further regulatory authority over Class V wells. Even as it moves to do so (58 Fed. Reg. 25,033 (1993)), there is no indication that injection for reuse will be a priority at the agency. Therefore, there is little immediate prospect of federally imposed standards for these types of projects. In any event, the limitations of constructing a regulatory scheme under the framework of the Safe Drinking Water Act should be noted. The act protects only drinking water sources, and aquifers used for other purposes are not necessarily protected. The emphasis of the act is on injection wells, so that, while they would be regulated under the act, surface spreading and soil-aquifer treatment (SAT) recharge of wastewater, for example, would not be. Finally, it is the violation of primary drinking water standards that is forbidden under the regulatory scheme (see, for example, 40 CFR § 144.12 (1992)). Many would argue for a higher degree of protection for an aquifer than wellhead compliance with these standards. A technology-based regulatory system could potentially result in better, albeit more costly, aquifer protection.

Aside from the UIC program, one must turn to state and local governments for laws governing ground water protection. State laws protecting ground water vary in what is regulated, the means of regulation, and the standards to which ground water is protected (National Research Council, 1986). A project that has a potentially detrimental effect on ground water may be subject to a permitting review to demonstrate that it would not cause ground water standards to be violated. These ground water standards are set by states without reference to a federal minimum level of protection, unlike most statutory schemes. States differ over whether they have a policy of nondegradation and over what degree of degradation of ground water is permitted. At the state and local level the control may vary depending on the source of the recharge water. The composition of the recharge water may determine the scrutiny that the project receives.

Arizona, Nevada, and Oregon provide examples of the different approaches taken by states. All three address ground water recharge statutorily and permits are required by each state (Ariz. Rev. Stat. Ann. § 45-652(B), § 45-672(B); Nev. Rev. Stat. Ann. § 534.015); Or. Rev. Stat. § 537.135 (1992)). However, the permit requirements are different in each state.

Of the three states, Arizona has the most extensive regulatory program. The permit applicant must show (1) the technical and financial capability of the project proponent; (2) the right to use the water for recharge or replenishment; (3) that the project is hydrologically possible; (4) that the project "will not cause unreasonable harm to land or other water users"; and (5) that the applicant has applied for an aquifer protection permit (Ariz. Rev. Stat. Ann. § 45-652(B), § 45-672(B)). The Aquifer Protection Permit program regulates point and nonpoint source discharges to aquifers. The program is designed to protect ground water, including "the protection of public health and the environment, preventing, mitigating, and remediating ground water contamination, regulating discharges to surface water and ground water and conserving ground water resources" (Redding and DuBois, 1990). One of the requirements of the aquifer protection permit is that the recharge water must not violate the Aquifer Water Quality Standards at the point of compliance (Ariz. Comp. Admin. r. and Regs. 18-11-401-408 (1992)). The point of compliance is the top of the aquifer, and credit can be given for treatment in the vadose zone. The Aquifer Water Quality

Standards encompass the federal primary drinking water standards and additional narrative standards. The narrative standards are meant to prevent discharges (or rather, recharges) which cause a pollutant to be present in an aquifer at levels that would endanger public health.

In Nevada, the state engineer oversees ground water recharge projects (Nev. Rev. Stat. Ann. § 534.250). Permits are required for recharge projects and, as in Arizona, the applicant must show (1) a legal right to use the water, (2) that land and other water users will not be harmed, and (3) that the project is technically, financially, and hydrologically feasible. However, specific ground water protection measures are not part of the Nevada statute. Instead, Nevada requires the operator of a recharge project to monitor the operation of the project and the "effect of the project on users of land and other water within the area of hydrological effect." Monitoring requirements are determined by the state engineer in cooperation with "all government entities which regulate or monitor. . . the quality of water."

Oregon has a short statutory provision addressing ground water recharge (Or. Rev. Stat. § 537.135 (1992). The appropriation of water for ground water recharge is recognized as a beneficial use. The Water Resources Commission is to develop standards that must be met for ground water recharge permits. The project must not be "prejudicial to the public interest" (Or. Rev. Stat. § 557.135 (1992)). A permit is also required to apply artificially stored water to beneficial use.

A corollary to ground water concerns arising from recharge projects should be noted. Projects that use a ground water basin for storage also acquire a practical interest in the protection of that basin from possible contamination from other sources. Sources of ground water contamination are ubiquitous, including underground storage tanks for petroleum products, landfills and hazardous waste facilities, septic tanks, pesticides and herbicides, agricultural fertilizer residues (nitrate), and virtually any other activity conducted on land. The desire to protect a ground water supply led to the Metropolitan Water District of Southern California to join an environmental group in challenging the expansion of a landfill in the San Gabriel Basin (Krautkraemer, 1991).

Use of Recharge Water

The use to which recharge water is put once it is recovered may also affect the regulations to which a project is subject. The Safe Drinking Water Act is intended to protect consumers of water through protection and regulation at the wellhead. Specific numeric parameters limit the presence of contaminants in the water supply. States can impose additional requirements on water to be reused for drinking. In Florida, for example, injection wells used to recharge ground water are subject to comprehensive regulation under the state administered UIC program (Fla. Admin. Code Ann. r. 17-28.011 to 17.610 (1985)) and related ground water regulations (see Fla. Admin. Code Ann. r. 17-600.540 (1985)). The Florida UIC program protects sources of drinking water and also protects the quality of aquifers used for other purposes. The regulations require that the injected fluid stay within the "injection zone," and "unapproved interchange of water between aquifers is prohibited." An injection zone is "a geological formation, group of formations, or part of formation receiving fluids directly through a well" (Fla. Admin. Code Ann. r. 17-28.120(39)).

Under the Florida UIC program, Class V wells are much more extensively regulated than under the federal program. Class V wells are divided into six categories, according to the "expected quality of the injected fluid," in order to determine what permitting, operating, and monitoring regulations are needed (Fla. Admin. Code Ann. r. 17-28.510). Group 2 wells are "recharge wells used to replenish, augment, or store water in an aquifer." In general, Class V wells must be so constructed that, at the point of discharge, water quality standards are not violated. The minimum criteria for ground water protection require that "all ground water shall at all places and at all times be free from domestic, industrial, agricultural, or other man-made non-thermal discharges of concentrations which" pose a threat, either carcinogenic, mutagenic, teratogenic, or toxic or which pose a threat to the public health, safety, or welfare (Fla. Admin. Code Ann. r. 17-520.400). In addition, the ground water shall be free from concentrations that are harmful to plants, animals, or organisms "which are native to the soil and responsible for treatment or stabilization of the discharge." Indigenous species of "significance to the aquatic community within surface waters affected by ground water at the point of contact with surface water" are also protected. Finally, the discharge must not create or constitute a nuisance or impair the reasonable and beneficial use of adjacent waters." The operation of Class V wells must not "cause or allow the movement of fluid containing any contaminant into underground sources of drinking water" if the presence of the contaminant will violate any primary drinking water standard (Fla. Admin. Code Ann. r. 17-28.610). The primary drinking water standards set maximum contaminant levels for organic and inorganic contaminants, turbidity, coliforms, and radionuclides (Fla Admin. Code Ann. r. 17-550.310). The primary drinking water standards may also serve as ground water quality standards.

To the degree recharge water is put to purposes other than drinking, such as landscape irrigation, there is no federal statute controlling these uses.

Environmental Consequences

Another regulatory mechanism with potential impact on ground water recharge is the National Environmental Policy Act (NEPA). NEPA was enacted to require federal agencies undertaking projects to consider the environmental consequences of "actions significantly affecting the quality of the human environment" (42 U.S.C. § 4332 (1988)). For projects for which an environmental impact statement is required, the act requires a far-reaching examination of consequences of projects and of alternatives. Public participation is also mandated. NEPA can also provide an opportunity for federal agencies to mitigate the environmental consequences of their projects, despite the lack of explicit regulatory language.

Some states review the environmental consequences of projects under so-called "little NEPAs." These statutes apply where the state or a local government is the proponent of a project and provide for similar far-reaching review of projects.

The Clean Water Act (Federal Water Pollution Control Act, 33 U.S.C. §§ 1251-1387 (1988)) may also come into play in some ground water recharge projects. For instance, a recharge project that occurs a streambed may require a federal permit to alter the streambed. Section 404 of the Clean Water Act mandates permits for "the discharge of dredged or fill materials" (33 U.S.C. § 1344 (1988)). The administrator of EPA can deny a permit if "the discharge of such materials into such area will have an unacceptable adverse effect on municipal water supplies, shellfish beds and fishery areas (including spawning and breeding areas), wildlife, or recreational areas" (Federal Water Pollution Control Act 33 U.S.C. § 1344 (1988)).

If the project to recharge water occurs in a streambed, it may also have to meet surface water quality standards if the source water is "discharged" to the stream. (This would occur, for example, if source water from stormwater or wastewater facilities were collected in some fashion and then discharged through a discrete conveyance to a stream.) A permit is not required under federal law for irrigation return flows that are discharged to a stream (33 U.S.C. § 1342 (l) (1988)).

California's Regulatory Controls: A Closer Look

California is the state where the greatest attention has been paid to the regulatory aspects of ground water recharge projects with treated wastewater. While, by statute, California actively encourages the reuse of wastewater for beneficial purposes, California's regulatory agencies have carefully scrutinized projects and reworked regulatory requirements. In this section, a brief history of this regulation development is presented.

History of Regulatory Activities in California

Although incidental and unplanned ground water recharge with municipal wastewater effluent has occurred for many years in California, it was not until the 1960s that the first large-scale, planned spreading operation began at Whittier Narrows in the Montebello Forebay area of Los Angeles County (Hartling, 1993). The injection of highly treated wastewater from Orange County Water District's

MILESTONES IN THE HISTORY OF GROUND WATER RECHARGE IN CALIFORNIA

1962 The first large-scale planned operation of ground water recharge was implemented when secondary effluent from the Whittier Narrows Water Reclamation Plant in Los Angeles County was spread in the Montebello Forebay area of the Central Groundwater Basin.

1973 The California Department of Health Services (DOHS) developed a position statement on the uses of reclaimed water involving ingestion, essentially placing a moratorium on new projects for ground water recharge.

1975 The State of California convened a Consulting Panel on the Health Aspects of Wastewater Reclamation for Groundwater Recharge to provide recommendations for research that would help DOHS establish statewide criteria for ground water recharge.

1976 DOHS developed draft regulations for ground water recharge that were subsequently used as guidelines.

1976 Ground water recharge by direct injection was initiated by the Orange County Water District to prevent saltwater intrusion.

1978 The Sanitation Districts of Los Angeles County (SDLAC) initiated a 5-year health effects study to investigate the health significance of using reclaimed water for ground water replenishment.

1978 Revisions made to the state's wastewater reclamation criteria included acceptance of the use of reclaimed water for recharge of potable aquifers by spreading. Specific requirements were to be determined on a case-by-case basis.

1984 SDLAC's health effects study was published and indicated that the spreading of reclaimed water into the Montebello Forebay had not resulted in observable adverse health effects on consumers of extracted ground water containing reclaimed water.

1986 The state of California appointed a Scientific Advisory Panel on Groundwater Recharge with Reclaimed Wastewater to provide information needed for the establishment of statewide criteria for ground water recharge.

1987 State regulatory agencies approved a 50 percent increase in the amount of reclaimed water that could be spread in the Montebello Forebay area.

1992 DOHS and the Santa Ana Regional Water Quality Control Board approved the injection of 100 percent reclaimed water by the Orange County Water District's Water Factory 21.

1993 DOHS released comprehensive draft regulations directed toward spreading and injection of reclaimed water into potable aquifers.

Water Factory 21 to create a seawater intrusion barrier, which is still the only project in California involving injection of reclaimed water, began with pilot testing in the late 1960s and became operational in 1976 (Argo and Cline, 1985). Increasing demands for water and economic and environmental concerns associated with new surface water development and large-scale water importation projects have given rise to expanded interest in the use of reclaimed water as a means of supplementing existing water supplies and meeting some of the future water needs of the state.

1973 Department of Health Position Statement

In the early 1970s, several water quality control plans ("basin plans") were developed under the direction of the State Water Resources Control Board (SWRCB). The basin plans identified 36 potential ground water recharge projects in the state. In 1973, the California State Department of Health Services (DOHS) prepared a position statement in response to proposals in the basin plans for augmentation of domestic water sources with reclaimed water. Three uses of reclaimed water were considered in the statement: (1) ground water recharge by surface spreading; (2) direct injection into an aquifer suitable for use as a domestic water source; and (3) direct discharge of reclaimed water into a domestic water supply system (California Department of Health Services, 1973).

The DOHS position statement recommended against direct discharge into a domestic water supply system and direct injection into aquifers used as a source of domestic water supply. It also stated that injection may be considered as an option and that injection for saline water intrusion barriers may be acceptable. In regard to surface spreading, the 1973 position statement concluded that surface spreading appears to have great potential as an option for ground water recharge even though information on health effects is uncertain. In regard to specific types of recharge projects, the recharge of small basins with large quantities was not to be recommended, but proposals for recharge of large basins with small amounts was to be considered, depending on details such as community well locations. The position statement essentially placed a moratorium on new projects for ground water recharge.

1975 Consulting Panel

As discussed in Chapter 4, in 1975 three state agencies—DOHS, SWRCB, and the Department of Water Resources (DWR)—jointly prepared a state-of-the-art report on the health aspects of water reclamation and reuse for ground water recharge (State of California, 1975). They convened a special panel, the Consulting Panel on the Health Aspects of Wastewater Reclamation for Groundwater Recharge, to recommend a program of research and demonstration projects to provide information to assist DOHS in the establishment of reclamation criteria for ground water recharge to augment potable water supplies. The panel was also asked to assist DWR and SWRCB in the planning and implementation of programs to encourage use of reclaimed water consistent with those criteria.

The panel, which confined its discussions to ground water recharge by surface spreading in order to better define the domain under consideration, concluded that uncertainties exist regarding potential health effects from the use of reclaimed water for ground water recharge, the principal concern being stable organic materials. In order to address the uncertainties, the panel recommended that comprehensive studies on the health effects of ground water recharge be initiated at existing projects and new demonstration projects established to gain field information under selected and controlled conditions. To provide a database for estimating health risk, research was recommended specifically in the areas of contaminant characterization, toxicology, and epidemiological studies of exposed populations (State of California, 1976).

In 1976, DOHS developed draft regulations for ground water recharge of reclaimed water by surface spreading (Crook, 1985). The proposed criteria were principally directed at the control of stable organic constituents. The minimum level of treatment specified in the draft regulations was conventional secondary treatment followed by carbon adsorption and percolation through at least 10 feet (3 m) of unsaturated soil. The proposed criteria included reclaimed water quality requirements, an effluent monitoring program, a minimum dilution requirement, and a minimum underground residence time of 1 year prior to ground water withdrawal. Other proposed requirements included detailed reports on hydrogeology and spreading operations, establishment of an industrial source control program, development of contingency plans, and implementation of a health monitoring program for the exposed population. The proposed regulations were not adopted as statewide criteria at that time but were used as guidelines for evaluating new ground water recharge projects.

In an attempt to answer some of the health-related issues associated with ground water recharge and implement the recommendations of the panel, the Sanitation Districts of Los Angeles County (SDLAC) initiated a health effects study in 1978 (Nellor et al., 1984). The focus of the study was the Whittier Narrows ground water recharge project in the Montebello Forebay area of Los Angeles County. The primary goal of the 5-year study was to develop a database that could be used to enable health and regulatory authorities to determine whether the use of reclaimed water for ground water replenishment at Whittier Narrows should be maintained at the present level, cut back, or expanded. A second goal of the study was to provide information to DOHS to use in establishing statewide reclamation criteria for recharge. (For more information, see Chapter 4.)

1986 Scientific Advisory Panel

As discussed in Chapter 4, the state of California commissioned a Scientific Advisory Panel on Groundwater Recharge with Reclaimed Water in 1986 to (1) define the health significance of using reclaimed water for ground water recharge to augment potable water supplies; (2) evaluate the benefits and risks associated with ground water recharge with reclaimed water; and (3) provide detailed background information needed for the establishment of statewide criteria for ground water recharge with reclaimed water (State of California, 1987a). The panel concluded that the Whittier Narrows ground water replenishment project should continue. Recharge via spreading was determined to be preferable to injection, and the panel concluded that available treatment processes can adequately remove organic constituents of concern. The panel also concluded that reclaimed water should be disinfected prior to injection or spreading; however, disinfection processes should not produce harmful by-products. It also concluded that all new projects should include prospective health surveillance of populations and biochemical testing of concentrates to determine whether substances likely to be harmful are present at low levels. State-of-the-art toxicology studies with animals were recommended for the purposes of risk evaluation. Finally, the panel recommended continued analytical chemistry investigation and monitoring to identify and quantify chemical constituents.

The Scientific Advisory Panel concurred with the health effects study findings and concluded that the risks, if any, associated with the Whittier Narrows recharge project were small and probably not dissimilar from those that could be hypothesized for commonly used surface waters. The panel tempered this conclusion with the statement that the results are "marginal or inconclusive" with regard to cancer because the exposure period was short in relation to the expected minimum 15-year latency period for chemically induced cancers (State of California, 1987b).

Water Reclamation Requirements Current and Proposed Regulations

While aspects of its regulatory development process have been protracted, California has developed a comprehensive approach to ground water recharge with treated municipal wastewater. A number of state agencies are involved in the regulation of ground water recharge projects. SWRCB establishes state water quality control policy and, along with the Regional Water Quality Control Boards (RWQCBs), is responsible for the protection of water quality (Cal. Water Code §§ 13140, 13523). DOHS establishes water reclamation and reuse criteria for each different use of reclaimed water (Cal. Water Code § 13521). The regional boards implement the water reclamation program requirements.

The RWQCBs, after consultation with DOHS, prescribe water reclamation requirements for projects proposing to use reclaimed wastewater. The RWQCBs issue permits authorizing the reclamation projects, and each permit includes the reclamation criteria relevant to the particular project (Cal. Water Code § 13523). The reclamation requirements must include, or conform with, the state criteria. The RWQCBs may impose regulations on "the person reclaiming the water, the user, or both."

The wastewater reclamation criteria found in the California Administrative Code, Title 22 §§ 60301-60329 (1994) define reclaimed water to be "water which, as a result of treatment of domestic waste water, is suitable for a direct beneficial use or a controlled use that would not otherwise occur." The requirements include current criteria for ground water recharge (Cal. Water Code Title 22 § 60320). A surface spreading project must have a permit and must meet the criteria. Water reclaimed by surface spreading "shall at all times be of a quality that fully protects public health." Under the current regulations, specific criteria are not established for ground water recharge by surface spreading, and projects are considered on a case-by-case basis. For each project, the RWQCB works with DOHS to establish permit requirements based on "all relevant aspects of each project, including the following factors: treatment provided, effluent quality and quantity, spreading area operations, residence time, and distance to withdrawal" (Cal. Water Code Title 22 § 60320).

Injection wells are addressed statutorily in section 13540 of the California Water Code. Injection wells may be used for ground water recharge if the regional board finds that "water quality considerations do not preclude" such activities and if DOHS, after a public hearing, determines that an injection well will not "impair the quality of water in the receiving aquifer" (Cal. Water Code § 13540). There are currently no regulations for injection wells in general, and requirements are imposed on a case-by-case basis.

DOHS formed a ground water recharge committee in 1988 to begin development of ground water recharge criteria. In a coordinated effort to address the regulatory needs associated with ground water recharge, DOHS, SWRCB, and DWR jointly developed a draft document entitled *Proposed Guidelines for Groundwater Recharge with Reclaimed Municipal Wastewater* in 1990 (State of California, 1990). The proposed guidelines were meant to help encourage and plan for the efficient use of the state's water resources and to increase water supply reliability by identifying the means for the safe use of treated municipal wastewater for ground water recharge. In addition, the proposed guidelines were to be a guide for the regional boards in establishing ground water quality objectives and water reclamation requirements. The guidelines were also to ensure that ground water recharge with reclaimed water, whether planned or incidental, would be regulated in a consistent manner. Finally, the guidelines were meant to assist in planning ground water recharge with reclaimed wastewater by providing criteria that detail what information is needed for review by regulatory agencies. The proposed guidelines discuss guiding principles, permitting procedures, and draft criteria for ground water recharge.

The proposed regulations have gone through several iterations (California Department of Health Services, 1993b). When finalized, the proposed criteria will replace the existing general regulations for ground water recharge in the DOHS wastewater reclamation criteria (Cal. Admin. Code Title 22 §§ 60301-60329 (1975)).

The proposed regulations address both surface spreading and injection projects, and are focused on indirect potable reuse of the recovered water. Treatment requirements and performance standards are proposed for each type of project. They also address water quality standards, recharge methods, operational controls, distance to withdrawal, time in the underground, and monitoring.

TABLE 5.2 Proposed Ground Water Recharge Criteria in California: Treatment Process and Site Requirements

	Project Category[a]			
	I	II	III	IV
Required treatment[b]				
Oxidation	x	x	x	x
Filtration	x	x		x
Disinfection	x	x	x	x
Organic Removal	x			x
Maximum allowable reclaimed water in extracted well water (%)	50	20	20	50
Depth to ground water (ft) at initial percolation rate of:				
<0.2 inches/minute	10	10	20	n.a.[c]
<0.3 inches/minute	20	20	50	n.a.
Minimum retention time underground (months)	6	6	12	12
Horizontal separation[d] (feet)	500	500	1,000	1,000

[a] Categories I, II, and III are for surface spreading projects; Category IV is for injection projects.
[b] X means that the treatment process is required.
[c] n.a. = not applicable.
[d] Distance from spreading area or injection well to nearest extraction well.

Source: California Department of Health Services, 1993a.

A summary of the proposed treatment process and site requirements is presented in Table 5.2.

The proposed regulations prescribe stringent microbiological and chemical constituent limits. For spreading operations, credit is given for chemical constituent removal and removal or inactivation of pathogenic microorganisms during percolation through the vadose zone, and the percolated water must be essentially pathogen-free and meet drinking water maximum contaminant levels after percolation.

The proposed regulations specify total organic carbon (TOC) as a surrogate for trace organic constituents that may be of concern. Although TOC is not a measure of specific organic compounds, it is considered to be a suitable measure of gross organic content of reclaimed water for the purpose of determining organic removal efficiency in practice. Based principally on the Scientific Advisory Panel report, DOHS concluded that extracted ground water should contain no more than 1 mg/l TOC of wastewater origin. This decision is reflected in maximum allowable TOC concentrations in the reclaimed water prior to spreading or injection.

Prior to adoption, the proposed regulations are subject to external review and any modifications that DOHS deems appropriate based on comments received. Hence, the proposed regulations may be substantially different from the ground water regulations that are ultimately adopted in California.

The wastewater reclamation criteria apply only to the reuse of domestic water. There are currently no regulations dealing with the use of stormwater runoff and irrigation return flow for ground water recharge, and such projects are dealt with on a case-by-case basis.

Other Relevant Laws

In addition to the laws specifically addressing ground water recharge, two other water protection laws, the Porter-Dolwig Ground Water Basin Protection Law (Cal. Water Code §12922) and the California Safe Drinking Water Act (Cal. Health and Safety Code § 4010), also relate to the use of reclaimed wastewater for ground water recharge.

The Ground Water Basin Protection Law seeks to ensure "the correction and prevention of irreparable damage to, or impaired use of, the ground water basins of this state caused by critical conditions of overdraft, depletion, seawater intrusion or degraded water quality" (Cal. Water Code, § 12922). The law applies to projects that, among other things, are used for reclamation of water used to "replenish, recharge or restore a ground water basin . . . when such basin is relied on as a source of public water supply" (Cal. Water Code § 12921.3). Under section 12923, projects may be reviewed by DWR and evaluated in terms of potential threats to the ground water and the project plans and design criteria may be revised in order to ensure that the ground water is protected (Cal. Water Code § 12923).

The California Safe Drinking Water Act is intended to ensure that the state has pure, wholesome, and potable drinking water (Cal Health and Safety Code § 4010). The regulations set forth primary and secondary drinking water standards, which are similar or more stringent than the federal standards, and are administered by DOHS. Operators of public water systems that extract ground water that is partially recharged wastewater must comply with the primary and secondary drinking water standards (Cal. Health and Safety Code § 4017).

INSTITUTIONAL ISSUES

Many institutional factors affect the viability of ground water recharge projects. In examining these influences, the fact of their mutability should be kept prominent: an uneconomic project can look feasible if subsidies are provided, or an unhelpful legal structure can be changed. Education programs can help shift public opinion.

Although the cost of alternative supplies is a critical factor, the existence of

---

WATER CONSERV II, ORANGE COUNTY FLORIDA

Water Conserv II is billed as the largest water reclamation project in the world that combines agricultural irrigation and rapid infiltration basins. The project uses water reclaimed from sewage to irrigate citrus groves and to recharge the Upper Floridan aquifer through rapid infiltration basins. It is a cooperative water conservation effort by the city of Orlando, Orange County, and the agricultural community.

The system is designed to produce 50 millon gallons of reclaimed water per day from two sewage treatment facilities. The reclaimed water is delivered to citrus grove owners under 20-year contracts for irrigation and frost protection purposes. What is not used for irrigation is routed to rapid infiltration basins for recharge of the Upper Floridan aquifer, the principal source of water for most of Florida. The system is designed to provide reclaimed water to 12,000 to 15,000 acres of citrus and to 2,000 acres of rapid infiltration basins.

This cooperative effort provides an excellent illustration of how collaborative arrangements can resolve both the sewage disposal and the water supply problems of an area. Agricultural users obtain reclaimed water at little or no cost, thereby increasing the profitability of their operations. Simultaneously, water supply from the Upper Floridan aquifer is enhanced both by the direct recharge via the rapid infiltration basins and by "in lieu recharge," which occurs because agricultural users reduce demand on the aquifer. The result is that there is greater availability of high-quality ground water for the potable water needs of the rapidly growing central Florida area at the same time that the agricultural demands of citrus growers are being met.

an entity with a motive to develop the water source is also critical. Thus, in a jurisdiction where ground water is developed for use by individuals (as in a rural agricultural area), and not by a central water district, ground water recharge may not occur because no individual has a sufficient stake to invest in its development. In southern California, in contrast, the Metropolitan Water District is a dominating entity with an interest in water conservation and the use of recharge basins (Tarlock, 1991).

While ground water recharge may be an appropriate means of augmenting water supplies, it can also be attractive because of a very different type of imperative. Disposal of wastewater can be expensive, especially with escalating pollution control requirements applied to discharges. These requirements have increased in cost and complexity, and are largely the result of federal and state level decision making. A "no discharge" operation may be appealing to the local government that is operating the wastewater disposal facility, regardless of the downstream ecological consequences.

Nonpoint source pollution has historically been exempt from pollution control regulation. There is some likelihood that pollution from nonpoint sources will become the focus of regulation in the future. Thus, for example, EPA's stormwater regulations (40 CFR § 122.26 (1992)) are now bringing stormwater runoff under regulatory control. These newly regulated dischargers face a choice not unlike that faced by industrial facilities 20 years ago: to what media shall the discharge be directed? If regulatory requirements for ground water disposal are less than those for surface water discharges, the use of artificial recharge projects may increase.

The statutory exemption of irrigation return flow from the Clean Water Act (33 U.S.C. § 1362(14)) has periodically been subject to criticism because of the constituents found in those discharges. If these sources are subjected to regulation by Congress, pressure might build for alternative disposal methods, including deliberate discharge to ground water.

Ground water recharge as a form of conjunctive use offers possible environmental advantages. Dams and reservoirs are criticized on environmental and economic grounds. Ground water storage can provide a means of achieving many of the same ends, but with fewer contentious aspects. Further, ground water storage projects can lessen demands on a river by allowing storage of peak flows, so these can be used at times of low flows.

The enthusiasm of a project developer is affected by the regulatory environment. While protracted reviews are no longer unexpected for large-scale projects, shifting regulatory targets can discourage project proponents. The content of regulations is frequently asserted to be less important than the certainty of regulations. For ground water recharge projects, the burden of being the first project in a state or jurisdiction can impose additional requirements on project proponents and discourage innovation.

The multiplicity of organizations involved in a typical recharge project might be thought to spell doom for any such project. This is a difficult proposition to test because it is harder to identify projects that never were initiated than to chronicle those that are now operational and therefore have overcome institutional barriers. Some research indicates that the mere multiplicity of actors is not an insurmountable barrier to these projects, when the need is great (Tarlock, 1991). This thesis is not immediately intuitively obvious, but examples abound where a multiplicity of powers successfully cooperate for common ends.

Because it can be expensive to research and develop new regulations, federal research into scientific and engineering issues and federal involvement in development of model statutes and guidelines can help states in their evaluation of proposed recharge projects. Needless to say, this research should reflect the substantial experience already garnered by the states and regions. The existence of this regulatory infrastructure can benefit those proposing projects and the public by providing clear guidance as to the jurisdiction's requirements. Whether the federal role should go beyond providing technical assistance is a matter of debate; given the innovation already evidenced by several states, and the variability in physical conditions among the regions of the nation, the merits of an expanded federal role should be evaluated as the pressure to utilize recharge grows. If states are stymied in the adoption of regulatory schemes, adopt regulations that result in health concerns, or ask for federal assistance, then federal regulation might be desirable.

Southern California's efforts to address its ground water problems have been the subject of considerable scholarly interest and are instructive in considering how regions have succeeded, despite a multiplicity of actors, in addressing public needs (see Ostrom, 1990). The challenges faced by the region are formidable, including the pressure of a growing population on water supplies, the expense of relying on imported surface water, a lack of controls on ground water mining, the danger of saltwater intrusion, and the sheer number of institutions with a role in water decisions. The solution adopted after many years of effort combined controls on ground water pumping, which were imposed through adjudication rather than state regulation, and the use of ground water recharge. Ground water recharge allowed the aquifers to be used for ground water storage and the replenishment of ground water to prevent saltwater intrusion. Significantly, the local water districts did not take the controlling law as static. In two instances, water producers from the districts pursued statutory changes, most notably securing passage of a statute that authorized them to form a joint district to control pumping, impose fees on pumping, and carry out recharge projects (Ostrom, 1990). What enables one region facing difficult water problems to overcome difficulties and arrive at a workable solution, while another stumbles into expensive and awkward solutions, is a query which confronts many areas of water management. Continuing research by social scientists will bring about a better understanding of the relevant institutional factors and how they interact.

> **RISK COMMUNICATION AND RISK PERCEPTION**
>
> In the past, risk communication was defined as a one-way transmission of expert knowledge to nonexperts. But this simple image has been replaced. Today, risk communication is seen as an interactive process of exchange of information and opinion among individuals, groups, and institutions. It involves multiple messages about the nature of risk and other messages, not strictly about risk, that express concerns, opinions, or reactions to risk messages or to legal and institutional arrangements for risk management. Risk communication is successful only to the extent that it raises the level of understanding of relevant issues or actions and satisfies those involved that they are adequately informed within the limits of available knowledge.
>
> "Risk perception," or how people judge and react to risk, deals with human values regarding attributes of hazards and benefits. Studies of risk perception, such as the studies cited in this report, typically present technologies, activities, or substances and ask people to consider the risks they feel each presents and to rate them. Analysis of such studies show that people's ratings are affected by certain attributes—such as the potential to harm large numbers of people at once, personal uncontrollability, dreaded effects, effects on children, reversibility, and perceived involuntariness of exposure—that make those hazards more serious to the public than hazards that lack those attributes. The fact that hazards differ dramatically in their qualitative aspects helps explain why certain technologies or activities, such as nuclear power, evoke more serious public opposition than others, such as motorcycle riding, that cause many more injuries and fatalities. This means that risk perception is value-laden. When lay and expert values differ, reducing different kinds of hazard to a common numerical rating (such as number of fatalities per year) and presenting comparisons only on that metric have great potential to produce misunderstanding and conflict and to engender mistrust of expertise.
>
> SOURCE: National Research Council, 1989.

PUBLIC ATTITUDES TOWARD THE USE OF RECLAIMED WATER

Water reclaimed from municipal wastewater or other sources of impaired quality holds the potential to be a significant source in water-short areas, but public opinion about such uses is a controlling factor (Bruvold, 1981). Indeed, the importance of the attitudes of the public cannot be discounted because the public is ultimately the recipient of the reclaimed water and it ultimately, albeit often indirectly, bears the burden of the costs of such operations. Thus, experts urge that the public be brought into the technical decision-making process early (Bruvold, 1976) and that communication of potential risks be presented in clear, plain language; be complete and be accurate (not distorting the risk or minimizing the existence of uncertainty); be oriented to the needs and concerns of the audience; and use risk comparisons cautiously (NRC, 1989). This type of effort can help develop the public's understanding of the issues involved in reclamation of wastewater and educate citizens so they act as informed decision makers.

People's attitudes about the reuse of reclaimed water depend on the source and the intended purpose of the reuse, and nonpotable reuse is more acceptable than potable reuse. When drinking water is at issue, indirect potable reuse is more acceptable to the public than direct potable reuse because the water is perceived to be cleansed as it flows in a river, lake, or aquifer (U.S. Environmental Protection Agency, 1992). However, in general the public does not favor potable reuse. For instance, in research designed to determine the attitudes of Californians toward the use of reclaimed water, Bruvold (1976) measured attitudes toward 25 uses of reclaimed water, ranging from high contact uses such as drinking and bathing to low-contact uses such as irrigating golf courses and road construction (Table 5.3). In general, the public accepts the use of reclaimed water for a variety of purposes, but not drinking or other high contact uses. Respondents opposed to various uses ranged from 56 percent who opposed drinking reclaimed water to only 1 percent who opposed the use of such water in road construction. Ground water recharge using reclaimed wastewater was opposed by 23 percent of those surveyed. That study and others (Table 5.3) are remarkably consistent in showing that a majority of about 55 percent do not want to use reclaimed wastewater for drinking, while about 45 percent say they would be willing to accept such reuse (Bruvold, 1975). Nonpotable uses such as toilet flushing and lawn and golf course irrigation were acceptable to most survey respondents.

Acceptance of reclaimed water for drinking and other high-contact uses is affected by public perception of the associated health risks: the public is concerned about the perceived quality of the water and whether it might serve to transmit pathogens, viruses, or harmful trace chemicals. Studies show relationships between certain beliefs and attitudes toward the reuse of water: people who believed that their water supply was polluted, that their area faced a water shortage, and that modern technology was available for purifying wastewater were consistently more favorable in attitude toward reuse than those who believed their water supply was not polluted, that their areas did not face shortage, and that modern technology could not reliably purify wastewater (Bruvold, 1975). Research investigating attitudes toward treatment and reuse options found that the public did not favor either (1) minimal treatment followed by ocean disposal or (2) high levels of advanced treatment and subsequent reuse for drinking. Instead, they preferred relatively high levels of treatment followed by a "middle" level of use (e.g., park and greenspace irrigation) (Bruvold, 1981).

It is important to note that these surveys indicate that the public was quite willing to accept many uses less intimate than ingestion. A sizable segment of those surveyed (ranging from 56 to 33 percent) did not oppose or was positive toward the use of reclaimed water even for drinking, thus leaving open the possibility of future acceptance of a wider variety of use for reclaimed water. For that

TABLE 5-3 Percentage of Respondents Opposed to Various Uses of Reclaimed Water In General Opinion Surveys

	Bruvold (1972) (N=972)	Stone & Kahle (1974) (N=1,000)	Kasperson et al. (1974) (N=400)
Drinking water	56	46	44
Food preparation in restaurants	56	-	-
Cooking in the home	55	38	42
Preparation of canned vegetables	54	38	42
Bathing in the home	37	22	-
Swimming	24	20	15
Pumping down special wells	23	-	-
Home laundry	23	-	15
Commercial laundry	22	16	-
Irrigation of dairy pasture	14	-	-
Irrigation of vegetable crops	14	-	16
Spreading on sandy areas	13	-	-
Vineyard irrigation	13	-	-
Orchard irrigation	10	-	-
Hay or alfalfa irrigation	8	9	-
Pleasure boating	7	14	13
Commercial air conditioning	7	-	-
Electronic plant process water	5	5	3
Home toilet flushing	4	5	-
Golf course hazard lakes	3	8	-
Residential lawn irrigation	3	6	-
Irrigation of recreation parks	3	-	-
Golf course irrigation	2	5	2
Irrigation of freeway greenbelts	1	-	-
Road construction	1	-	-

Note: dash indicates that particular use was not included in survey.

Source: Bruvold (1987).

to happen, Bruvold (1976) suggested that "those who wish to demonstrate that reclaimed water is of high quality should initiate highly visible, well publicized demonstrations using reclaimed water for low-contact purposes not likely to be controversial. Such innovations would give technical experts, health officials, and the lay public experiential and scientific evidence that modern technology can provide water that is reliably of high quality in every respect. If these demonstration efforts are successful we will be a long way ahead in developing public acceptance for reclaimed water that might eventually include intimate personal use and consumption."

Olson et al. (1979) (N=244)	Bruvold (1981) (N=140)	Milliken & Lohman (1983) (N=399)	Lohman & Milliken (1985) (N=403)
54	58	63	67
57	-	-	-
52	-	55	55
52	-	55	55
37	-	40	38
25	-	-	-
40	-	-	-
19	-	24	30
18	-	-	-
15	-	-	-
15	21	7	9
27	-	-	-
15	-	-	-
10	-	-	-
8	-	-	-
5	-	-	-
9	-	-	-
12	-	-	-
7	-	3	4
5	8	-	-
6	5	1	3
5	4	-	-
3	4	-	-
5	-	-	-
4	-	-	-

SUMMARY

The future of ground water recharge using waters of impaired quality will be crucially affected by the economic, legal, and institutional setting. Indeed, the institutional barriers may prove to be more problematic than any remaining technical constraints. From an economic perspective, aquifer recharge with waters of impaired quality may be more attractive in the future because of the increasing scarcity of new sources of surface water. Also, increasingly stringent wastewater discharge regulations may make the incremental costs of rendering wastewater fit for potable or nonpotable uses quite modest in comparison with the costs of other new sources.

The economic feasibility of recharge with waters of impaired quality will vary from situation to situation. Reclaimed waters will be attractive from an economic viewpoint whenever they are the least-cost source of supplemental water. The costs of treatment over and above what is required to meet wastewater discharge standards will be particularly important. However, costs will also be quite sensitive to the distance that reclaimed waters have to be transported for spreading or injection and to the techniques used for spreading or injection. Economic feasibility will also depend on the benefits that the recharge waters ultimately yield. As long as users are willing to defray the full costs of recharge, there is *prima facie* evidence that the benefits will outweigh the costs.

From both an economic and a legal perspective, the need to define rights to both source waters and product waters is paramount. Failure to define rights clearly by itself makes recharge with waters of impaired quality far less attractive than it might otherwise be.

From a strictly legal standpoint, the central question is how to formulate policy to protect public health, the public good, and the environment, while not imposing inappropriate or unnecessarily burdensome controls on recharge facilities. Most existing laws focus on the need to protect ground water quality, but there is also a body of law directed at the use of recharge water.

State laws governing recharge are highly variable. California and Arizona have detailed and comprehensive sets of laws and regulations, but many other states have not addressed this regulatory problem or have done so inadequately. Although there are federal laws that govern certain aspects of the recharge process, the federal government has not exercised strong leadership in developing appropriate institutions to govern wastewater recharge. For the most part, the development of such institutions has been highly decentralized, and this approach may not prove workable in the future.

REFERENCES

Argo, D. G., and N. M. Cline. 1985. Groundwater recharge operations at Water Factory 21, Orange County, California. In Artificial Recharge of Groundwater, T. Asano, ed. Boston, Mass.: Butterworth.

Brown, G. M., Jr., and R. Deacon. 1972. Economic optimization in a single cell aquifer. Water Resour. Res. 8(3):557-564.

Bruvold, W. H. 1975. Human perception and evaluation of water quality. Crit. Rev. in Environ. Cont. 5(2):153-231.

Bruvold, W. H. 1976. Using reclaimed water: Public attitudes and governmental policy. In Public Affairs Report, Bulletin of the Institute of Governmental Studies, Vol. 17, No. 3. University of California, Berkeley.

Bruvold, W. H. 1981. Community evaluation of adopted uses of reclaimed water. Water Resour. Res. 17:487-490.

Bruvold, W. H. 1987. Public evaluation of salient water reuse options. In Proceedings of Water

Reuse Symposium IV, Denver, Col. Aug. 2-7, 1987. Denver, Col.: Am. Water Works Assoc. Res. Found.

Burness, H. S., and W. E. Martin. 1988. Management of a tributary aquifer. Water Resour. Res. 24(5):1339-1344.

Burt, O. R. 1970. Groundwater storage control under institutional restrictions. Water Water Resour. Res. 24(5):1540-1548.

California Department of Health. 1973. Position on Basin Plans for Reclaimed Water Uses Involving Ingestion. Water Sanitation Section, California Department of Health. Berkeley, Calif.

California Department of Health Services. 1993a. Draft Groundwater Recharge Regulations. Office of Drinking Water Technical Operations Section, California Department of Health Services. Berkeley, Calif.

California Department of Health Services. 1993b. Proposed Regulation. California Code of Regulations, Title 22, Division 4, Chapter 3. Office of Drinking Water, California Department of Health Services. Berkeley, Calif.

Crook, J. 1985. Regulatory approach to groundwater recharge with reclaimed domestic wastewater. Pp. 1673-1694 in Proceedings of Water Reuse Symposium III, August 26-31, 1984, San Diego, California. Denver, Col.: Amer. Water Works Res. Found.

Cummings, R. G. 1970. Some extensions of the economic theory of exhaustible resources. West. J. Econ. 7(3):201-210.

Cummings, R. G. 1971. Optimum exploitation of groundwater reserves with saltwater intrusion. Water Resour. Res. 7(6):1415-1424.

Gisser, M. 1983. Groundwater: Focusing on the real issue. J. Polit. Econ. 91(4):1001-1027.

Hartling, E. C. 1993. Impacts of the Montebello Forebay Groundwater Recharge Project. Bull. Calif. Water Pollut. Contr. Assoc. 29(3):14-26.

Ingram, H. M., D. E. Mann, G. D. Weatherford, and H. J. Cortner. 1984. Guidelines for improved institutional analysis in water resources planning. Water Resour. Res. 20(3):323-334.

Krautkraemer, J. 1991. Pros and Cons of Ground Water Regulation. Pp. 151-153 in Proceedings: Changing Practices in Ground Water Management—The Pros and Cons of Regulation. Rep. No. 77. Water Resources Center. University of California.

McGinnis, M. 1990. Creating a "new" class of water-regulation of municipal effluent, *Arizona Public Service Co. v. Long*, 160 Ariz. 429, 773 P.2d 988 (1989), Arizona State L.J. 22(4):987-1002.

National Research Council. 1986. Ground Water Quality Protection: State and Local Strategies. Washington, D.C.: National Academy Press.

National Research Council. 1989. Improving Risk Communication. Washington, D.C.:National Academy Press.

Nellor, M. H., R. B. Baird, and J. R. Smyth. 1984. Health Effects Study—Final Report. NTIS No. PB-84191-568. County Sanitation Districts of Los Angeles County, Whittier, Calif.

Ostrom, E. 1990. Governing the Commons: The Evolution of Institutions for Collective Action. New York: Cambridge University Press.

Orange County Water District/County Sanitation Districts of Orange County. 1993. A Joint Water Recycling Strategy for Achieving Water Supply Independence. Draft report dated Oct. 8, 1993. p. 19.

Provencher, B., and O. R. Burt. 1993. The externalities associated with common property exploitation of groundwater. J. Environ. Econ. Manage. 24(2):139-158.

Redding, M. B., and J. F. DuBois. 1990. Arizona's Aquifer Protection Permit Program: A regulatory approach to protecting groundwater. In Proceedings of the Underground Injection Practices Council Research Foundation Conferences/Symposia at Monterey, Calif. December 10-12, 1990.

State of California. 1975. A State-of-the Art Review of Health Aspects of Wastewater Reclamation for Groundwater Recharge. Prepared by the California State Water Resources Control Board,

Department of Water Resources, and Department of Health. Published by the Department of Water Resources, Sacramento, Calif.

State of California. 1976. Report of the Consulting Panel on Health Aspects of Wastewater Reclamation for Groundwater Recharge. Prepared by the California State Water Resources Control Board. Published by the Department of Water Resources, Sacramento, Calif.

State of California. 1987a. The Porter-Cologne Water Quality Control Act. California State Water Resources Control Board, Sacramento, Calif.

State of California. 1987b. Report of the Scientific Advisory Panel on Groundwater Recharge with Reclaimed Wastewater. Prepared for State of California Water Resources Control Board, Department of Water Resources, and Department of Health Services, Sacramento, Calif. State of California.

State of California. 1990. Proposed Guidelines for Groundwater Recharge with Reclaimed Municipal Wastewater. Prepared by the California State Water Resources Control Board, Department of Water Resources, and Department of Health Services, Sacramento, Calif. State of California.

Tarlock, A. D. 1991. Coordinating Water Resources in the Federal System: The Groundwater-Surface Water Connection, A-118. Washington, D.C.: U.S. Advisory Commission on Intergovernmental Relations.

U.S. Environmental Protection Agency, U.S. Agency for International Development. 1992. Manual: Guidelines for Water Reuse. EPA/625/R-92/004. Office of Water, U.S. Environmental Protection Agency.

Vaux, H. J., Jr. 1985. Economic aspects of groundwater recharge. Pp. 703-718 in Artificial Recharge of Groundwater. T. Asano, ed. Boston, Mass.: Butterworth

Warren, J. P., L. L. Jones, R. D. Lacewell and W. L. Griffin. 1975. External costs of land subsidence in the Houston Baytown area. Am. J. Agric. Econ. 57(4):450-455.

Western States Water Council. 1990. Ground Water Recharge Projects in the Western United States: Economic Efficiency, Financial Feasibility and Legal/Institutional Issues. Denver, Col.: U.S. Department of Interior.

6

Selected Artificial Recharge Projects

Although much can be learned from discussing the general characteristics of ground water recharge technologies, the knowledge becomes most useful when seen in light of actual examples. Examples provide an opportunity to see how theory translates into on-the-ground activity. This chapter provides brief descriptions of existing ground water recharge projects. The examples were selected to illustrate the common techniques used, show a variety of the purposes for which recharge is planned, and give concrete examples of the problems such projects sometimes face. The seven sites discussed are

- Water Factory 21, Orange County, California
- Montebello Forebay, California
- Phoenix, Arizona
- El Paso, Texas
- Long Island, New York
- Orlando, Florida
- The Dan Region, Israel

These examples are illustrative and brief, and the committee did not attempt to make recommendations from these site-specific cases. Instead, the committee hopes that these descriptions will show that artificial ground water recharge is not a "technology of the future" but rather something in use today in relatively diverse settings. These examples show that properly planned and operated artificial recharge projects can increase our water management options and flexibility.

The Water Factory 21 project in Orange County, California, is the first injection project involving highly treated municipal wastewater. The injected water provides a barrier to seawater intrusion, but also enters potable water supply aquifers. The project has been in operation since 1976 and has provided significant data on the capability and reliability of advanced wastewater treatment processes to remove microbiological and chemical constituents, ground water quality, and monitoring techniques.

Another California example is the Montebello Forebay project in south-central Los Angeles County. This project demonstrates indirect potable reuse via surface spreading of reclaimed water. The Montebello Forebay project has been in operation since 1962 and has been the subject of extensive research to investigate health-related issues.

The Phoenix, Arizona, example illustrates the extensive research undertaken to demonstrate the capability of soil-aquifer treatment (SAT) to treat relatively low quality treated municipal wastewater to levels acceptable for many non-potable applications upon extraction.

The El Paso, Texas, project is the first injection project in the United States where the sole intent of the project is to augment the potable water supply aquifer using reclaimed municipal wastewater. It is a relatively new project and will provide important data as it builds an operational history.

The Long Island, New York, example demonstrates the effectiveness of artificial recharge in a more urbanized, eastern setting, where climate and water availability are significantly different than in the West. Stormwater runoff is recharged into infiltration basins to replenish the ground water withdrawn for use by Long Island residents, thereby also helping to retard seawater intrusion into the aquifers that provide the primary source of drinking water for the area.

Another eastern project, the stormwater drainage wells in Orlando, Florida, is included to illustrate another approach to using excess stormwater runoff for artificial recharge, thus helping to solve a wastewater disposal problem as well as a water supply problem.

Finally, one international example is provided. The Dan Region project in Israel provides information on a large-scale recharge operation that incorporates SAT of treated municipal wastewater and subsequent extraction of the water for extensive agricultural irrigation. The project is well documented and has been in operation for almost 20 years.

WATER FACTORY 21, ORANGE COUNTY, CALIFORNIA

The Orange County Water District (OCWD) was formed by a special act of the California legislature in 1933 for the purpose of protecting the Orange County ground water basin. In 1955, OCWD was given the added responsibility of water management. Early in its history, OCWD secured the right to all water in

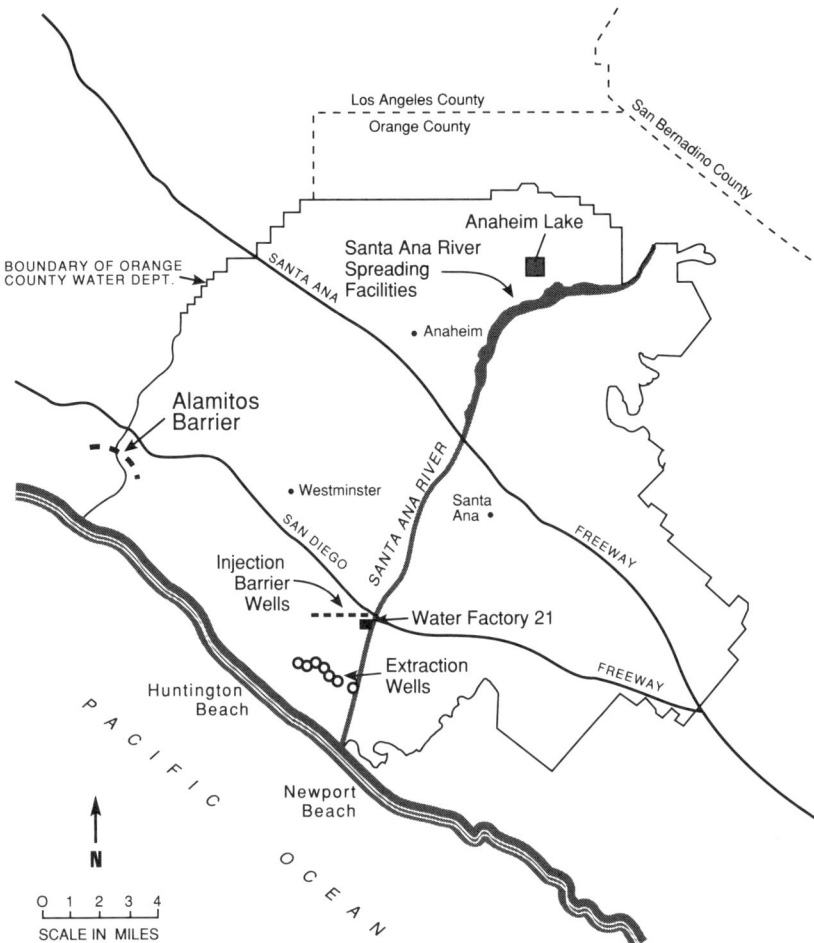

FIGURE 6.1 Orange County ground water recharge facilities and saltwater intrusion barriers.

the Santa Ana River. Over 3 million acre-feet of the river's flow has been captured to recharge the Orange County ground water basin. In addition, more that 2.5 million acre-feet of water imported from northern California and the Colorado River has been recharged (Orange County Water District, 1991). The location of OCWD and its recharge facilities is shown in Figure 6.1.

The Orange County ground water basin is the depositional plain of the Santa Ana River. The principal features of the region are surrounding hills and a broad, poorly drained alluvial plain with alternating gaps and minor hill systems along the coast. A major fault system parallels the coastline, which apparently seals the basin from the sea at deeper levels. However, in several gaps along the ocean front there is hydraulic continuity between seawater and ground water in the upper 45 to 60 m (150 to 200 ft) of recent alluvial fill (Argo and Cline, 1985).

The aquifers in the area are composed of fine- to coarse-grained sand, separated by silt and clay layers (aquicludes and aquitards). The Talbert aquifer, the principal zone of production in the area, is of recent age and overlies Pleistocene deposits within the gap created by the Santa Ana River. The Talbert aquifer is the only aquifer in direct contact with the Pacific Ocean. The three lower zones of local production are subject to intrusion by virtue of their contact with the Talbert aquifer.

The base of the freshwater-bearing sediments is more than 1,200 m (4,000 ft) deep in some inland locations but rises to a depth of 60 m (200 ft) along the coast, where seawater intrusion has occurred. Seawater intrusion was first observed in municipal wells during the 1930s as a consequence of basin overdraft. Overdrafting of the ground water continued into the 1950s. Overpumping of the ground water resulted in seawater intrusion as far as 5.6 km (3.5 miles) inland from the Pacific Ocean by the 1960s. Although OCWD prevented further intrusion through percolation of large amounts of imported water in the forebay area of the ground water basin, the need for a coastal barrier system was obvious.

OCWD began pilot studies in 1965 to determine the feasibility of using effluent from an advanced wastewater treatment (AWT) facility as injection water in a hydraulic barrier system to prevent the encroachment of saltwater into potable water supply aquifers. Construction of an AWT facility known as Water Factory 21 was started in 1972 in Fountain Valley, and injection of the treated municipal wastewater into the ground began in 1976.

Water Factory 21 receives activated sludge secondary effluent from the adjacent County Sanitation Districts of Orange County (CSDOC) and has a design capacity of 15 million gallons per day (mgd). Water Factory 21 has the following unit processes: lime clarification for removal of suspended solids, heavy metals, and dissolved minerals; air stripping for removal of ammonia and volatile organic compounds; recarbonation for pH control; mixed-media filtration for removal of suspended solids; activated carbon adsorption for removal of dissolved organic compounds; reverse osmosis (RO) for demineralization and removal of other constituents; and chlorination for disinfection and algae control. The current operation mode is shown in Figure 6.2.

Because California rules require that total dissolved solids cannot exceed 500 milligrams per liter (mg/l) prior to injection, RO is used to demineralize up to 5 mgd of the wastewater used for injection. The feed water to the RO plant is effluent from the mixed-media filters. Effluent from granular activated carbon adsorption columns is disinfected and blended with RO water. Activated carbon is regenerated on site in a multiple-hearth furnace. Solids from the settling basins are incinerated in a multiple-hearth furnace from which lime is recovered

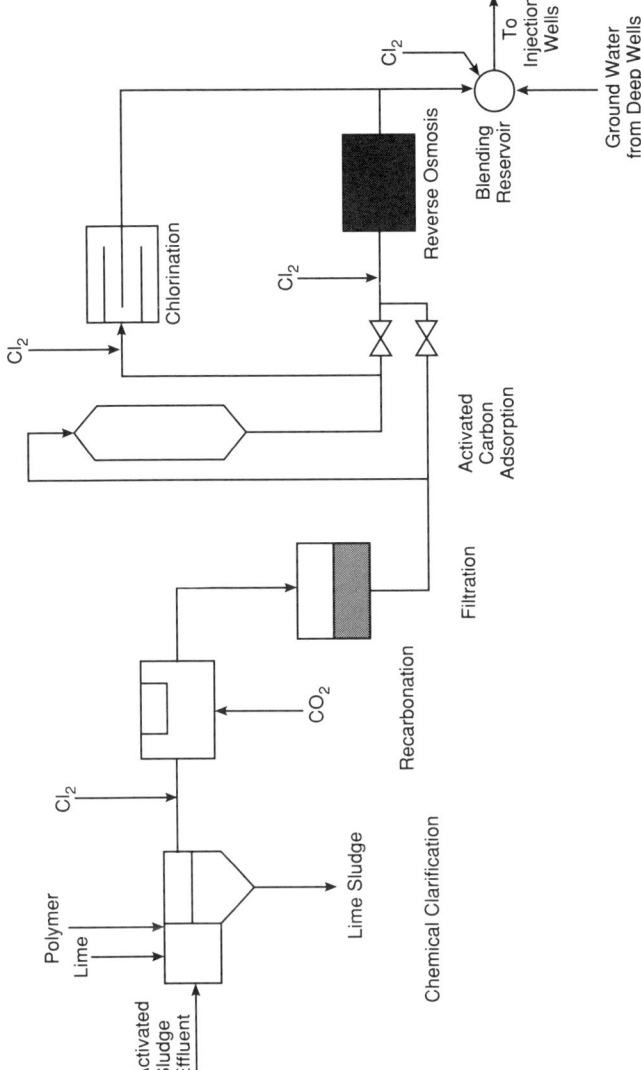

FIGURE 6.2 Flow schematic for Orange County Water District Water Factory 21.

and reused in the chemical clarifier. Brine from the RO process is pumped to the CSDOC facilities for ocean disposal. Reclaimed water produced at Water Factory 21 is injected into a series of 23 multicasing wells providing 81 individual injection points into 4 aquifers. The resulting seawater intrusion barrier is known as the Talbert injection barrier (Argo and Cline, 1985). A schematic of a typical injection well is shown in Figure 6.3. The wells are located at 183-m (600-ft) intervals in a city street approximately 5.6 km (3.5 miles) inland from the Pacific Ocean. Each well has the capacity to inject 450 gallons per minute (gpm). They vary in depth from 27 m (90 ft) to 130 m (430 ft). There are 7 extraction wells located between the injection wells and the coast. At the present time, the ground water is maintaining a positive hydraulic gradient toward the ocean, and the extraction wells are not in use. Prior to injection, the product water is blended 2:1 with deep well water from an aquifer not subject to contamination. The blended water is chlorinated in a blending reservoir before it is injected into the ground. Depending on conditions, the injected water flows toward the ocean forming a seawater barrier, inland to augment the potable ground water supply, or in both directions. On average, well over 50 percent of the injected water flows inland to augment the potable water supply.

The AWT processes at Water Factory 21 reliably produce a high-quality water. No total coliform organisms were detected in any of 161 samples of blended injection water tested during 1990 (Wesner, 1991). A virus monitoring program conducted from 1975 to 1982 demonstrated to the satisfaction of the state and county health agencies that Water Factory 21 effluent is essentially free of measurable levels of viruses (McCarty et al., 1982). The average turbidity of filter effluent was 0.20 nephelometric turbidity units (NTU) and did not exceed 1.0 NTU at any time during 1990. The average chemical oxygen demand (COD) and total organic carbon (TOC) concentrations for the year were 8 mg/l and 2.8 mg/l, respectively (Wesner, 1991). The effectiveness of the RO process in the removal of inorganic constituents at Water Factory 21 is indicated in Table 2-10 in Chapter 2. The concentrations of priority organic pollutants at various steps in the treatment train are presented in Table 2.11 in Chapter 2.

In 1992, the California Department of Health Services removed a restriction that required blending reclaimed water not of sewage origin prior to injection. Hence, OCWD is considering phasing out use of deep well water for blending and inject 100 percent reclaimed water. In addition, ground water studies indicate that approximately 25 mgd of injected water is needed to fully protect against seawater intrusion at the Talbert Gap, and consideration is being given to increasing the amount of water produced at Water Factory 21 by 5 to 10 mgd in future years.

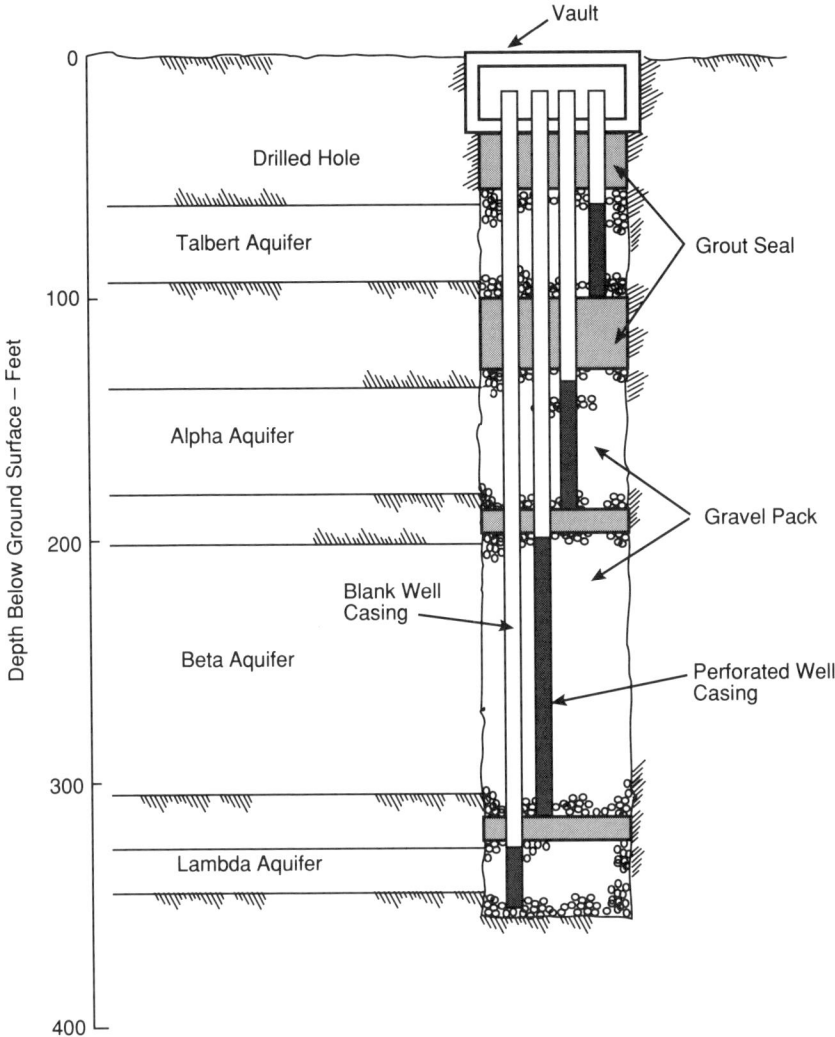

FIGURE 6.3 Typical reclaimed water injection well.

MONTEBELLO FOREBAY GROUND WATER RECHARGE PROJECT, LOS ANGELES, CALIFORNIA

Ground water is an integral component of southern California's water resources. Artificial recharge of aquifers is practiced to augment replenishment of ground water basins in several locations, including the Montebello Forebay area of south-central Los Angeles County. Waters used to recharge via surface spreading include local stormwater runoff, imported surface water (Colorado River water and State Project water), and reclaimed municipal wastewater.

Imported surface water is not always available for recharge during the summer months when demands by domestic water systems are at their peak. In addition, the availability of imported surface water is likely to be severely limited in the future. When the Central Arizona Project is completed in the mid-1990s, California's allotment of imported Colorado River water could be reduced by as much as 600,000 acre-feet/year. Also, the state of California may limit deliveries from the State Project, which supplies water to southern California from the Feather River/Sacramento Delta in northern California.

Reclaimed water has been used as a source of ground water replenishment in the Montebello Forebay area since 1962. At that time, approximately 12,000 acre-feet/year of disinfected activated sludge secondary effluent from the Sanitation Districts of Los Angeles County (LACSD) Whittier Narrows Water Reclamation Plant (WRP) was spread in the Montebello Forebay area of the Central Groundwater Basin, which is the main body of ground water underlying the greater Los Angeles metropolitan area. The basin has an estimated usable storage capacity of 780,000 acre-feet. In 1973, the San Jose Creek WRP was placed in service and also supplied secondary effluent for recharge. In addition, effluent from the Pomona WRP that is not reused for other purposes is discharged into San Jose Creek, a tributary of the San Gabriel River, which ultimately becomes a source of recharge water in the Montebello Forebay. The use of effluent from the Pomona WRP is expected to decrease as the reclaimed water becomes more fully used for irrigation and industrial applications in the Pomona area.

The water reclamation plants were originally built as secondary treatment facilities; however, body contact recreational activities in the receiving waters dictated that additional public health protection measures be taken. In the late 1970s all three reclamation plants were upgraded to provide tertiary treatment via dual media filtration (for the Whittier Narrows and San Jose Creek WRPs) or activated carbon filtration (for the Pomona WRP), and chlorination/dechlorination (Nellor et al., 1984). The activated carbon filters at the Pomona WRP have since been converted to dual-media filters.

The Montebello Forebay ground water recharge project is a cooperative effort. LACSD collects and treats municipal wastewater and monitors the effluent quality. The replenishment program is operated by the Los Angeles County Department of Public Works (LADPW), while overall management of the ground water basin is administered by the Water Replenishment District of Southern California (WRDSC). LADPW constructed special spreading areas designed to increase the indigenous percolation capacity by modifying the San Gabriel River channel and constructing off-stream spreading basins, ranging in size from 4 acres to 20 acres, adjacent to the Rio Hondo and San Gabriel rivers. The Rio

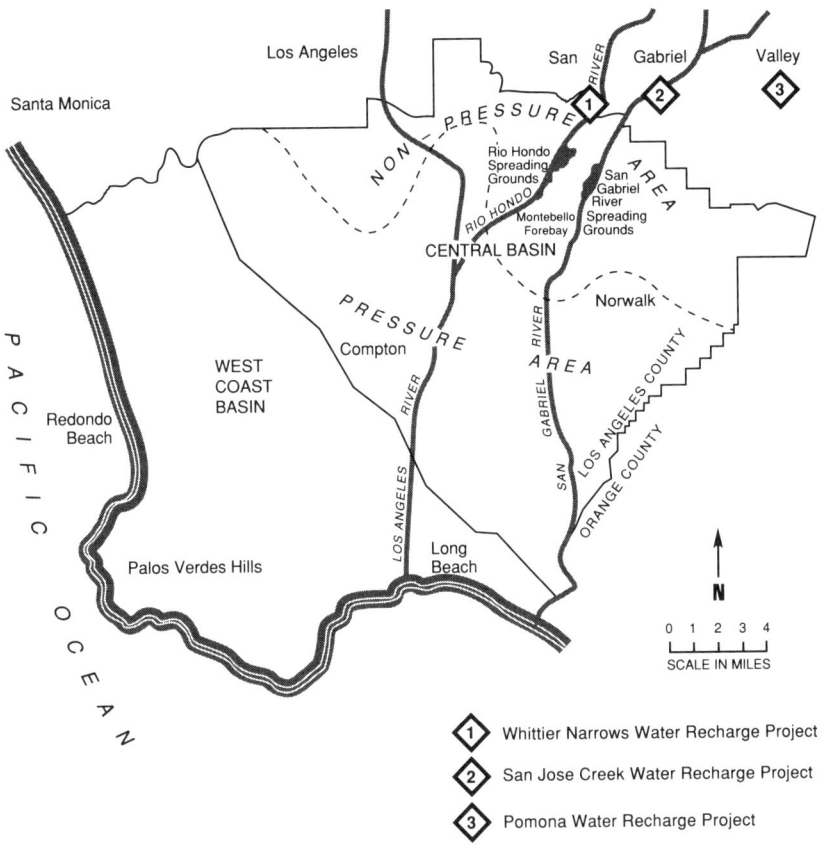

FIGURE 6.4 Montebello Forebay ground water recharge facilities.

Hondo spreading basins have 427 acres available for spreading. The San Gabriel River spreading basins occupy 224 acres, which include approximately 133 acres in an unlined section of San Gabriel River. The locations of the spreading basins and water reclamation plants are shown in Figure 6.4.

Under normal operating conditions, the basins are rotated through a 21-day cycle consisting of (1) a 7-day flooding period during which the basins are filled to maintain a constant 1.2 m (4-ft) depth; (2) a 7-day draining period during which flow to the basins is terminated and the basins are allowed to drain; and (3) a 7-day drying period during which the basins are allowed to dry out thoroughly. This wetting/drying operation serves several purposes, including maintenance of aerobic conditions in the upper soil strata.

In the aftermath of the 1976-1977 drought, there was considerable pressure to more fully use reclaimed water supplies in southern California, particularly for ground water recharge. However, concerns by the California Department of Health Services (DOHS) over potential health effects of using reclaimed water to replenish potable water supplies caused a moratorium on planned expansions. In an attempt to answer some of the health-related issues associated with ground water recharge, a Health Effects Study was initiated in 1978 (Nellor et al., 1984). The focus of the study, conducted by LACSD, was the Montebello Forebay ground water recharge project. At the time the study was conducted, the annual amount of reclaimed water spread and recharged averaged 26,500 acre-feet/year, which was 16 percent of the total inflow to the ground water basin, with no more than 32,700 acre-feet of reclaimed water spread in any given year. The percentage of reclaimed water in the ground water supply was estimated to range from 0 to 23 percent on an annual basis, and 0 to 11 percent on a long-term (1962 to 1977) basis.

The primary goal of the 5-year $1.4 million study was to develop a database which could be used to enable health and regulatory authorities to determine whether the use of reclaimed water for ground water replenishment in the Montebello Forebay should be maintained at the then-current level, cut back, or expanded. A wide range of research was undertaken, including (1) water quality characterizations of ground water, reclaimed water, and other recharge sources in terms of their microbiological and inorganic chemical content; (2) toxicological and chemical studies of ground water, reclaimed water, and other recharge sources to isolate and identify health-significant organic constituents; (3) percolation studies to evaluate the efficacy of soil in attenuating inorganic and organic chemicals in reclaimed water; (4) hydrogeological studies to determine the movement of reclaimed water through ground water and the relative contribution of reclaimed water to municipal water supplies; and (5) epidemiological studies of populations ingesting recovered water to determine if their health characteristics differed significantly from a demographically similar control population.

The results of the Health Effects Study indicated that the risks associated with the three sources of recharge water (i.e., imported water, stormwater, and reclaimed water), were not significantly different and the historical proportion of reclaimed water used for replenishment had no measurable impact on either ground water quality or the health of the population ingesting the water (Nellor et al., 1984). The epidemiological study findings are weakened somewhat by recognition that the minimum observed latency period for human cancers that have been linked to chemical agents is about 15 years. Because of the relatively short time period that ground water containing a substantial proportion of reclaimed water had been consumed, it is unlikely that examination of cancer mortality rates would have detected an effect, if present, of exposure to reclaimed water.

Based on the results of the Health Effects Study and recommendations of a state-sponsored Scientific Advisory Panel (State of California, 1987), authorization was given by the Los Angeles Regional Water Quality Control Board (LARWQCB) and DOHS in 1987 to increase the annual quantity of reclaimed water used for replenishment by approximately 50 percent to 50,000 acre-feet/year over a period of 3 years, contingent upon the evaluation of data generated by an expanded monitoring program. Other requirements limited the total quantity of reclaimed water spread in any year to 50 percent of the total inflow to the basin and stipulated that the reclaimed water must meet all drinking water maximum contaminant levels and action levels (i.e., concentrations of contaminants in drinking water at which adverse health effects would not be anticipated to occur, based on an annual running average). Approval also was contingent on demonstration that there was no measurable increase in organic chemical contaminants in the ground water as the result of using reclaimed water for recharge.

Since the initial authorization, three increments of 7,300 acre-feet/year have been implemented, increasing the quantity of reclaimed water for ground water recharge to 50,000 acre-feet/year, or approximately 30 percent of the total inflow to the Montebello Forebay. In 1991, the LARWQCB revised permit conditions to allow recharge of up to 60,000 acre-feet of reclaimed water in any one year as long as the running 3-year average does not exceed 150,000 acre-feet. This allowed for greater flexibility in spreading operations.

The Montebello Forebay ground water recharge project includes extensive sampling and analysis of reclaimed water from the Whittier Narrows, San Jose Creek, and Pomona WRPs with similar monitoring of six shallow monitoring wells within the confines of the spreading grounds, 20 production wells in and around the spreading grounds, and ground water both upgradient and downgradient of the spreading grounds. The results of this combined monitoring program indicate that there has been no degradation of the ground water quality in terms of total dissolved solids, nitrogen, trace organics, heavy metals, or microorganisms (Hartling, 1993). Sampling and analysis of reclaimed water from each of the WRPs indicate that the WRPs consistently produce reclaimed water that does not contain measurable levels of viruses, contains less than 2.2 total coliform organisms/100 ml, and has an average turbidity of less than 2 NTU. Tables 2.8 and 2.9 in Chapter 2 provide water quality data from the three water reclamation plants that provide reclaimed water for recharge in the Montebello Forebay.

In addition to providing a much-needed source of water for recharge, the use of reclaimed water is attractive from an economic standpoint. In 1992, WRDSC purchased reclaimed water from the Whittier Narrows WRP for $7per/acre-foot and reclaimed water from the San Jose Creek WRP for $11.56 per/acre-foot. Reclaimed water from the Pomona WRP that is not reused for other purposes, approximately 2,000 acre-feet, is captured for ground water recharge at no cost. The cost of the reclaimed water compares favorably to the seasonal storage rate of $130 per acre-foot for imported water purchased from the Metropolitan Water District of Southern California in 1992 (Hartling, 1993). The seasonal storage rate is normally in effect during the wet, winter months and is expected to increase substantially in future years. The use of reclaimed water in place of imported water has saved an estimated $44 million since recharge using reclaimed water began in 1962 (Hartling, 1993).

Additional research conducted since the completion of the Health Effects Study has included an evaluation of the efficiency of the LACSD's full-scale carbon filters for removing mutagenicity as determined by the *Salmonella* microsome assay (Baird, 1987). This work indicates that average removals of 80 percent could be achieved based on a 10-minute empty bed contact time and that the effects of chlorine disinfection on mutagenic activity vary significantly. These results suggest that chlorine can oxidize (deactivate) some types of mutagens but also can react with available organic matter to create more mutagens in a given sample.

Ongoing research has focused on the development of a ground water tracer suitable for characterizing the movement of reclaimed water in ground water basins. The study has thus far evaluated a series of alkyl pyridone sulfonate (APS) compounds and several fluorocarbon compounds in the laboratory to measure the degree of adsorption of these compounds on soils, their ability to withstand biodegradation under aerobic and anaerobic conditions, and their ability to withstand photodecomposition. Volatility studies and biological assays have been conducted to determine the potential of the tracer compounds to elicit acute toxicity or mutagenicity. The laboratory phase of study has been completed, and the second phase of study will verify the laboratory results under actual field conditions (R. B. Baird, personal communication, 1992).

In 1993, research was initiated to provide comparative, supplemental data for the Health Effects Study findings. Similar toxicological and chemical procedures are being used to characterize any changes in reclaimed water or ground water quality that might have occurred since the Health Effects Study samples were originally collected for evaluation. Additionally, the researchers will use current techniques to learn more about the characteristics of compounds in mutagenic fractions, thereby providing a better understanding of the origins and health significance of these compounds as well as the alternatives available for their removal (Sloss, 1993).

PHOENIX, ARIZONA, PROJECTS

The city of Phoenix and other municipalities in the Salt River Valley of Arizona are interested in renovating part of their treated municipal wastewater by SAT so that it can be used for unrestricted irrigation and stored underground for eventual potable use. There are two major sewage treatment plants in the Phoenix area: the 91st Avenue treatment plant (activated sludge, chlorination, capacity about 119,000 million gallons/day (mgd)) and the 23rd Avenue treatment plant (activated sludge, chlorination, capacity about 40 mgd. The feasibil-

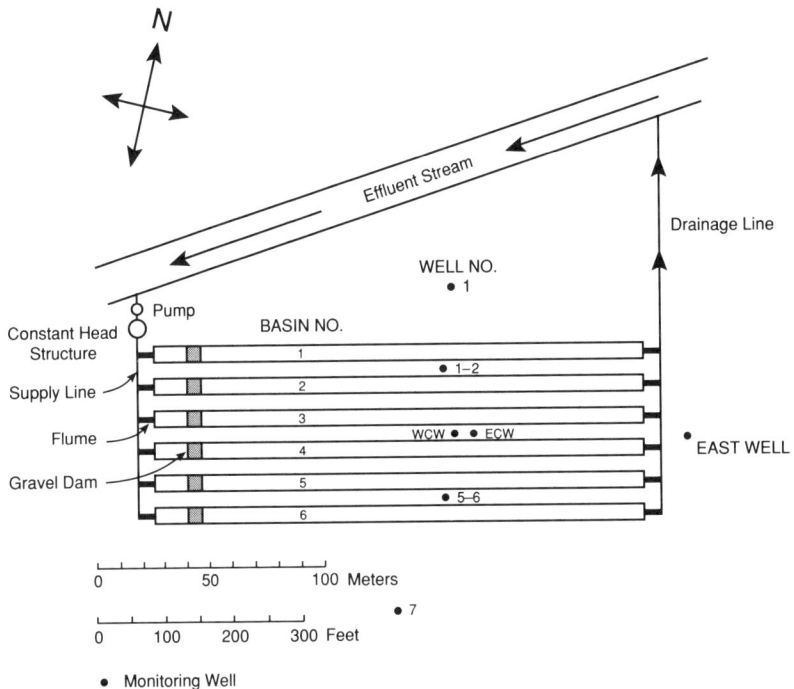

FIGURE 6.5 Schematic of Flushing Meadows project showing infiltration basins and monitoring wells.

ity of SAT in the Phoenix area was studied with two experimental systems, a small test project installed in 1967 below the 91st Avenue treatment plant, and a larger, demonstration project installed in 1975 below the 23rd Avenue treatment plant. The latter could be part of a future operational project that would have a basin area of 119 acres and a projected capacity of about 73×10^9 gal/year. Both projects were in the normally dry Salt River bed.

The project below the 91st Avenue treatment plant, known as the Flushing Meadows project (Bouwer et al., 1974a,b, 1980), was an experimental project that consisted of six parallel, long, narrow infiltration basins of about 0.32 acres each (Figure 6.5) in the Salt River floodplain. The soil consisted of about 1 m (3 ft) of loamy sand underlain by sand and gravel layers. The ground water table was at a depth of around 3 m (9 ft). Monitoring wells 2 to 9 m (7 to 30 ft) deep were installed at various points between the basins and away from the basins. This configuration made it possible to sample renovated wastewater from the aquifer both while it was still below the basins and after it had moved laterally for some distance through the aquifer.

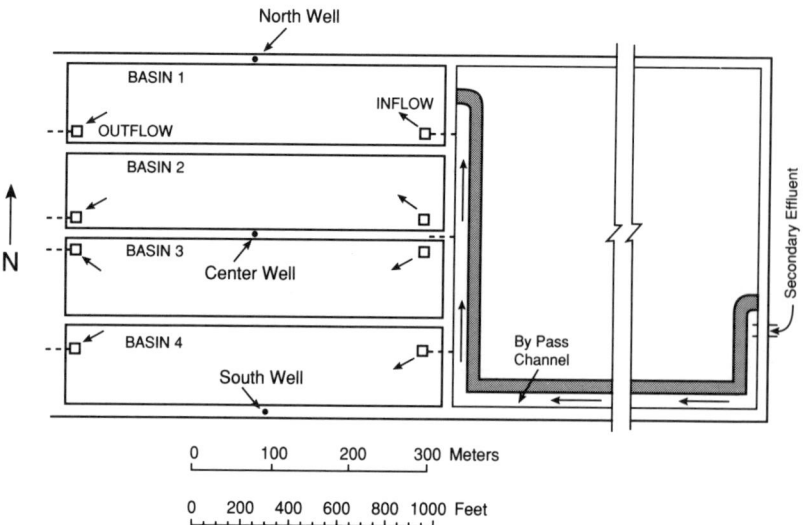

FIGURE 6.6 Sketch of the 23rd Avenue project with the 16-ha lagoon (left) that was split into 4 infiltration basins for the demonstration project, and the 32-ha lagoon (right) that can be split into 9 infiltration basins to increase the capacity of the system to that of plant outflow.

The second project was the 23rd Avenue project (Bouwer and Rice, 1984). This was a demonstration and possible future operational project on the north side of the Salt River bed. It consisted of an old 39.5-acre lagoon split lengthwise into four infiltration basins of 9.9 acres each (Figure 6.6). The soil lacked the loamy sand top layer of the Flushing Meadows project. Thus, the soil profile consisted mostly of sand and gravel layers. At this site, the ground water table was much deeper and ranged between 5 and 25 m (15 and 75 ft) depth (mostly around 15 m (45 ft)) for the study period. Monitoring wells to sample renovated wastewater were installed in the center of the project at depths of 18, 24, and 30 m (59, 79, and 98 ft), and on the north and south sides of the basin complex at depths of 222 m (728 ft) (Figure 6.6). In addition, a large production well (capacity about 2.6 mgd) was drilled in the center of the project with the casing perforated from 30 to 54 m (98 to 177 ft) depth.

Infiltration Rates

The inundation schedule typically was 9 days flooding/12 days drying at the Flushing Meadows project and 14 days flooding/14 days drying at the 23rd Avenue project. Water depths in the basins were about 15 to 20 cm (6 to 8 inches). During flooding, infiltration rates typically were between 0.3 and 0.6

m/day (1 and 2 ft/day), yielding a total infiltration or hydraulic loading of between 60 and 120 m/year (197 and 394 ft/year) for Flushing Meadows and about 100 m/year (330 ft/year) for the 23rd Avenue project.

For the 23rd Avenue project, the effluent from the treatment plant initially flowed through a 79-acre lagoon before it entered the infiltration basins. Heavy growth of algae, particularly the unicellular *Carteria klebsii*, in the lagoon caused soil clogging in the infiltration basins, especially in the summer. In addition to forming a "filter cake" on the bottom of the infiltration basins, the photosynthesis of the algae removed carbon dioxide from the wastewater, which raised the pH and, in turn, caused precipitation of calcium carbonate, which further aggravated the soil clogging. Because of this clogging, hydraulic loading rates initially averaged only 21 m/year (69 ft/year). Algae growth and resulting soil clogging were prevented by building a bypass canal around the lagoon (Figure 6.6), reducing the detention time of the effluent from a few days in the lagoon to about one-half hour in the canal. Also, the water depth in the basins was reduced from about 100 to 20 cm (39 to 7.9 inches). This increased the rate of turnover of the water in the basins and reduced growth of suspended algae. It also reduced compaction of the clogging layer, thus yielding higher infiltration rates (Bouwer and Rice, 1989). After the bypass canal was put into operation, hydraulic loading rates for the infiltration basins were almost 100 m/year (330 ft/yr), or approximately 5 times higher than they were before.

At a hydraulic loading rate of 100 m/year (330 ft/yr), 1 acre of infiltration basin can handle $330 \times 325,900 = 10^8$ gal/year of wastewater. Thus, the 40 mgd of effluent from the 23rd Avenue wastewater treatment plant would require 144 acres of infiltration basins. Almost all of this area could be obtained by also converting the 79-acre lagoon east of the present infiltration system into infiltration basins. This would give a total basin area of about 118 acres, which could handle about 126 million gallons/year of effluent. The wells for pumping the renovated water from the aquifer would be located on the centerline through the project. At a capacity of 2.6 mgd per well, 12 wells would be needed to pump renovated water out of the aquifer at the rate at which it infiltrates as wastewater in the basins, thus creating an equilibrium situation (Bouwer and Chase, 1984).

Removal of Contaminants by SAT

For both recharge projects, most of the quality improvement of the wastewater occurred in the vadose zone (i.e., the zone between soil surface and the ground water table). The quality improvements are summarized below (for additional details, see H. Bouwer et al. (1974b, 1980), Gilbert et al. (1976), and E. J. Bouwer et al. (1984)):

Suspended Solids

The suspended solids content of the renovated water at the Flushing Meadows project was less than 1 mg/l, compared to a range of 10-70 mg/l for the secondary effluent (average was 32 mg/l in 1977). For the 23rd Avenue project, it averaged about 1 mg/l for the large production well. Most of these solids probably were fine aquifer particles that entered the well through the perforations in the casing. The suspended solids content of the secondary effluent at the 23rd Avenue project averaged about 11 mg/l.

Total Dissolved Solids

The total salt or dissolved solids content (TDS) of the water increased slightly as it moved through the SAT system (from 750 to 790 mg/l at the 23rd Avenue project). Evaporation from the basins (including from the soil during drying) should increase the TDS content by about 2 percent. The rest of the increase was probably due to mobilization of calcium carbonate due to a pH drop from 8 to 7 as the wastewater moved through the vadose zone.

Nitrogen

At the Flushing Meadows project, nitrogen removal by SAT was about 30 percent at maximum hydraulic loading (100 to 120 m/year (330 to 390 ft/year)). This increased to 65 percent when the loading rate was reduced to about 70 m/year (230 ft/year) by using 9-day flooding and 12-day drying cycles and by reducing the water depths in the basins from 0.3 to 0.15 m (1 to 0.5 ft). The form and concentration of nitrogen in the renovated water sampled from the aquifer below the basins were slow to respond to the reduction in hydraulic loading (Bouwer et al., 1980). In the tenth year of operation (1977), the renovated water contained 2.8 mg/l of ammonium nitrogen, 6.25 mg/l nitrate nitrogen, and 0.58 mg/l organic nitrogen, for a total nitrogen content of 9.6 mg/l. This was 65 percent less than the total nitrogen of the secondary treated wastewater, which averaged 27.4 mg/l (20.7 mg/l as ammonium) in that year.

At the 23rd Avenue project, the total nitrogen content in the treated wastewater averaged about 18 mg/l, of which 16 mg/l was as ammonium. The 2-week flooding and drying cycles must have been conducive to denitrification in the vadose zone because the total nitrogen content of the renovated water from the large center well averaged 5.6 mg/l, of which 5.3 mg/l was as nitrate, 0.1 mg/l as ammonium, 0.1 mg/l as organic nitrogen, and 0.02 mg/l as nitrite. The nitrogen removal thus was about 70 percent. This removal was the same whether or not the secondary treated wastewater was chlorinated in the treatment plant, indicating that the low residual chlorine content of the treated wastewater by the time it infiltrated into the ground apparently had no effect on the nitrogen transformations in the soil.

The flooding and drying sequence that maximizes denitrification in the vadose zone depends on various factors and must be evaluated for each particular system. Pertinent factors include the ammonium and carbon contents of the treated wastewater entering the soil, infiltration rates, cation exchange capacity of soil, exchangeable ammonium percentage, depth of oxygen penetration in the soil during drying, and temperature. The combined laboratory and field data from the Flushing Meadows experiments showed that to achieve high nitrogen removal percentages, the amount of ammonium nitrogen applied during flooding must be balanced against the amount of oxygen entering the soil during drying. Flooding periods must be long enough to develop anaerobic conditions in the soil. Infiltration rates must be controlled to the appropriate level for the particular wastewater, soil, and climate at a given site. Most of the nitrogen transformations in the Flushing Meadows studies occurred in the upper 50 cm (20 inches) of the vadose zone.

Phosphate

Phosphate removal increased with increasing distance of underground movement of the treated wastewater. After 3 m (9.8 ft) of downward movement through the vadose zone and 6 m (19.8 ft) mostly downward through the aquifer, phosphate removal at the Flushing Meadows project was about 40 percent at high hydraulic loading and 80 percent at reduced hydraulic loading. Additional lateral movement of 60 m (197 ft) through the aquifer increased the removal to 95 percent (i.e., to a concentration of 0.51 mg/l phosphate phosphorus versus 7.9 mg/l in the effluent). After ten years of operation and a total infiltration of 754 m (2470 ft) of secondary treated wastewater, there were no signs of a decrease in phosphate removal.

At the 23rd Avenue project, phosphate phosphorus concentrations in the last few years of the research averaged 5.5 mg/l for the secondary treated wastewater going into the ground and 0.37 mg/l for the renovated water pumped from the center well. The shallower wells showed a higher phosphate content, indicating that precipitation of phosphate again continued in the aquifer. For example, renovated water sampled from the 22-m- (72-ft-) deep north well showed phosphate phosphorus concentrations that averaged 1.5 mg/l, or about 4 times more than for the deeper center well. Most of the phosphate removal probably was due to precipitation of calcium phosphate.

Fluoride

Fluoride removal paralleled phosphate removal, indicating precipitation as calcium fluoride. At the Flushing Meadows project, fluoride concentrations in

1977 were 2.08 mg/l for the effluent, 1.66 mg/l for the water after it had moved 3 m (9.8 ft) through the vadose zone and 3 to 6 m (9.8 to 19.7 ft) through the aquifer, and 0.95 mg/l after it had moved an additional 30 m through the aquifer. At the 23rd Avenue project, fluoride concentrations averaged 1.22 mg/l in the secondary effluent and 0.7 mg/l in the renovated water from the center well.

Boron

Boron was not removed in the vadose zone and the aquifer of the Flushing Meadows and 23rd Avenue projects and was present at concentrations of 0.5 to 0.7 mg/l in both treated wastewater and renovated water. The lack of boron removal was due to insufficient amounts of clay in the vadose zone and aquifer.

Metals

At the Flushing Meadows project, movement of the secondary treated wastewater through 3 m (9.8 ft) of vadose zone and 6 m (19.7 ft) of aquifer reduced zinc from 193 to 35 µg/l, copper from 123 to 16 µg/l, cadmium from 7.7 to 7.2 µg/l, and lead from 82 to 66 µg/l (Bouwer et al., 1974b). Cadmium thus appeared to be the most mobile metal. More recent (about 1990) analyses showed much lower metal concentrations in the treated municipal wastewater. This is probably due to the use of better analytical equipment (atomic absorption with graphite furnace) and better control of industrial waste discharges into the sewer system. More recent concentrations were 36 µg/l for zinc, 8 µg/l for copper, 0.1 µg/l for cadmium, and 2 µg/l for lead, which would give much lower metal concentrations in the water after SAT.

Fecal Coliforms

The secondary treated municipal wastewater at the Flushing Meadows project was not chlorinated and contained 10^5 to 10^6 fecal coliforms per 100 ml. Most of these were removed in the top meter of the vadose zone. Some penetrated to the aquifer, however, especially when a new flooding period was started. The deeper penetration of fecal coliforms at the beginning of a flooding period was attributed to less straining of bacteria at the soil surface because the clogging layer had not yet developed. Also, since the activity of native soil bacteria at the end of a drying period can be expected to be lower because the input of nutrients was stopped, there probably was a less antagonistic environment for the fecal coliforms in the soil when flooding was resumed. Fecal coliform concentrations in the water after 3 m (9.8 ft) of travel through the vadose zone and 6 m (19.7 ft) through the aquifer were 10 to 500 per 100 ml when the renovated water consisted of water that had infiltrated at the beginning of a flooding period, and between 0 and 1 per 100 ml after continued flooding.

Additional lateral movement of about 100 m (330 ft) through the aquifer was necessary to produce renovated water in which fecal coliforms were undetectable in 100 ml at all times.

At the 23rd Avenue project, fecal coliform concentrations in the secondary sewage effluent entering the infiltration basins were 10,000 per 100 ml prior to November 1980 when the effluent was not yet chlorinated in the treatment plant and was first passed through the 79-acre lagoon. This concentration increased to 1.8×10^6 per 100 ml when the nonchlorinated effluent was bypassed around the lagoon and flowed directly into the infiltration basins. It then decreased to 3,500 per 100 ml after the effluent was chlorinated in the treatment plant and still bypassed around the lagoon. The corresponding fecal coliform concentrations in the water pumped from the large center well from a depth of 30 to 54 m (98 to 177 ft) for these periods averaged 2.3, 22, and 0.27 per 100 ml, respectively, with ranges of 0 to 40, 0 to 160, and 0 to 3 per 100 ml, respectively. Higher fecal coliform concentrations were observed in the renovated water from the shallower wells, especially when the fecal coliform concentration of the infiltrating effluent was 1.8×10^6/per 100 ml. At that time, water from the 18-m- (59-ft-) deep well showed coliform peaks after a new flooding period was started that regularly exceeded 1,000/per 100 ml and at one time even reached 17,000/per 100 ml. Thus, a considerable number of fecal coliforms passed through the vadose zone. However, chlorination of the effluent and resulting reduction of the fecal coliform concentration to 3,500/per 100 ml prior to infiltration, and additional movement of the water through the aquifer to the center well produced renovated water that was essentially free from fecal coliforms.

Viruses

At the Flushing Meadows project, the viral concentrations of nonchlorinated secondary effluent averaged 2,118 plaque forming units (pfu) per 100 liter (average of six bimonthly samples taken for 1 year). Identified viruses included polio, echo, coxsackie, and reoviruses. No viruses could be detected in renovated water sampled after 3 m (9.8 ft) of movement through the vadose zone and 3 to 6 m (9.8 to 19.7 ft) of movement through the aquifer (Gilbert et al., 1976). At the 23rd Avenue project, viral concentrations in the renovated water from the center well averaged 1.3 pfu per 100 liter before chlorination of the secondary effluent, and 0 pfu per 100 liter after chlorination of the secondary effluent. The combined effects of chlorination and SAT thus apparently resulted in almost complete removal of the viruses, considering that the virus assays were done on 800 to 2,000 liter samples of the water after SAT.

Organic Carbon

At the Flushing Meadows project, the biochemical oxygen demand (BOD) of the effluent after moving 3 m (9.8 ft) through the vadose zone and 6 m (19.7 ft) through the aquifer was essentially zero (too small to determine), indicating that almost all biodegradable carbon was mineralized. However, the renovated water still contained about 5 mg/l total organic carbon (TOC), as compared to 10 to 20 mg/l of TOC in the secondary effluent.

At the 23rd Avenue project, the TOC concentration of the secondary effluent averaged 12 mg/l where it entered the infiltration basins and 14 mg/l at the opposite ends of the basins. This increase probably was due to biological activity in the water as it moved through the basins. The renovated water from the 18-m (59-ft) well (intake about 5 m (16 ft) below the bottom of the vadose zone) had a TOC content of 3.2 mg/l and that from the center well (which pumped from 30- to 54-m (98- to 177-ft) depth) had a TOC content of 1.9 mg/l, indicating further removal of organic carbon as the water moved through the aquifer. The TOC removal in the SAT system was the same before and after chlorination of the secondary treated wastewater, indicating that chlorination had no effect on the microbiological processes in the soil.

The concentration of organic carbon in the renovated water (1.9 mg/l) was higher than the 0.2 to 0.7 mg/l typically found in unpolluted ground water. The latter concentrations are mostly due to humic substances such as fulvic and humic acids (Thurman, 1979). The renovated sewage water from the SAT process thus could contain a number of synthetic organic compounds, some of which could be carcinogenic or otherwise toxic, or trihalomethane (THM) precursors.

Trace Organic Compounds

The nature and concentration of trace organics in the secondary sewage treated wastewater and in the renovated water from the various wells of the 23rd Avenue project were determined by Stanford University's Environmental Engineering and Science Section, using gas chromatography and mass spectrometry. The studies were carried out for 2 months with nonchlorinated effluent, and then for 3 months with chlorinated effluent. As could be expected, the results showed a wide variety of organic compounds, including priority pollutants (many in concentrations on the order of micrograms per liter (see E. J. Bouwer et al., 1984; and H. Bouwer and Rice, 1984).

The chlorination had only a minor effect on the type and concentration of organic compounds in the treated municipal wastewater. Of the volatile organic compounds, 30 to 70 percent were lost by volatilization from the infiltration basins. Soil percolation removed 50 to 99 percent of the nonhalogenated organic compounds, probably mostly by microbial decomposition (Table 6.1). Concentrations of halogenated organic compounds decreased to a lesser extent with passage through the soil and aquifer (Table 6.2). Thus, halogenated organic compounds (including the aliphatic compounds chloroform, carbon tetrachloride, trichloroethylene, and 1,1,1-trichloroethane, and the aromatic di- and tri-

TABLE 6.1 Percentage Decrease in Concentration of Nonhalogenated Hydrocarbons During Passage Through Unsaturated Zone

	Without Chlorination		With Chlorination	
	Geometric Mean Concentration of Secondary Effluent (27 samples) (µg/l)	Average Decrease in Renovated Water from 18-m Well (6 samples) (%)	Geometric Mean Concentration of Secondary Effluent (27 samples) (µg/l)	Average Decrease in Renovated Water from 18-m Well (6 samples) (%)
Aliphatic hydrocarbons				
5-(2-Methylpropyl) nonanes	0.35	>94	0.57	>96
2,2,5-Trimethylhexane	0.11	>82	0.18	>89
6-Methyl-5-nonene-4-one	0.41	93[a]	0.94	98[a]
2,2,3-Trimethylnonane	0.21	76[a]	0.25	>92
2,3,7-Trimethyloctane	0.12	50[a]	0.27	>93
Aromatic hydrocarbons				
o-Xylene	0.45	67[a]	0.50	88[a]
m-Xylene	0.76	78[a]	1.00	98[a]
p-Xylene	0.17	53[a]	0.12	92[a]
C3-benezene isomer	0.56	84[a]	0.34	>94
C3-benezene isomer	0.48	85[a]	0.53	96[a]
Styrene	0.26	>92	0.58	98[a]
2,2,4-Trimethylbenzene	0.80	78[a]	1.04	96[a]
Ethylbenzene	0.19	53[a]	0.15	67
Naphthalene	0.22	68[a]	0.63	91[a]
Phenanthrene	0.10	80	0.10	90
Diethyl phthalate	19	20	10	90

[a]Level of significance for the difference between basin and well concentrations based on a t test comparison is less than or equal to 0.1.

Source: E. J. Bouwer et al., 1984.

chlorobenzenes and chlorophenols) were more mobile and refractory in the underground environment than the nonhalogenated compounds, which included the aliphatic nonanes, hexanes, and octanes, and the aromatic xylenes, C_3-benzenes, styrene, phenanthrene, and diethyl phthalate.

Other Organic Micropollutants

In addition to the aliphatic and aromatic compounds mentioned, other compounds tentatively identified in organic extracts of the samples of treated municipal wastewater and renovated water using gas chromatography and mass spectrometry were fatty acids, resin acids, clofibric acid, alkylphenol poly-

TABLE 6.2 Percentage Decrease in Concentration of Halogenated Organic Substances During Passage Through Unsaturated Zone

	Without Chlorination		With Chlorination	
	Geometric Mean Concentration of Secondary Effluent (27 samples) (µg/l)	Average Decrease in Renovated Water from 18-m Well (6 samples) (%)	Geometric Mean Concentration of Secondary Effluent (27 samples) (µg/l)	Average Decrease in Renovated Water from 18-m Well (6 samples) (%)
Chlorinated aliphatic hydrocarbons				
Chloroform	2.72	61[b]	3.46	88[b]
1,1,1-Trichloroethane	2.94	34	1.41	84[b]
Carbon Tetrachloride	0.12	0	0.12	42[b]
Bromodichloromethane	—[a]		0.26	> 62[b]
Trichloroethylene	0.91	−180[b]	0.39	−267[b]
Dibromochloromethane	—		0.23	> 57
Tetrachloromethane	2.63	−97[b]	1.69	81[b]
Bromoform	—		0.08	> 10
Chlorinated aromatics				
o-Dichlorobenzene	3.52	25	2.40	10
m-Dichlorobenzene	0.79	58[b]	0.38	5
p-Dichlorobenzene	2.25	33[b]	1.81	10
1,2,4-Trichlorobenzene	0.19	42[b]	0.38	71[b]
Trichlorophenol	0.01	0	0.02	0
Pentachlorophenol	0.02	0	0.04	0
Pentachloroanisole	0.43	−150	0.18	

[a]— Dash indicates not detected.
[b]Level of significance for the difference between basin and well concentrations based on a t test comparison is less than or equal to 0.1.

Source: E. J. Bouwer et al., 1984.

ethoxylate carboxylic acids (APECs), trimethylbenzene sulfonic acid, steroids, n-alkanes, caffeine, diazinon, alkylphenol polyethoxylates (APEs), and trialkylphosphates. Several of the compounds were detected only in the secondary effluent and not in the renovated water. A few others—diazinon, clofibric acid, and tributyl- phosphate—decreased in concentration with soil passage, but were still detected in the renovated water. The APEs appeared to undergo rather complex transformations during filtration through the soil. They appeared to be completely removed with soil percolation when nonchlorinated effluent was infiltrated; when chlorinated effluent was used, however, two isomers were found following soil filtration, while others were removed.

These studies show that SAT is effective in reducing concentrations of a number of synthetic organic compounds in treated municipal wastewater, but that the renovated water still contains a wide spectrum of organic compounds, albeit at very low concentrations. Thus, while the renovated water is suitable for unrestricted irrigation and recreation, recycling it for drinking would require additional treatment, such as membrane filtration to remove the remaining organic compounds. The water would also have to be disinfected. Treatment of renovated municipal wastewater from an SAT system for potable use would, however, be much more effective and cheaper than treatment of municipal wastewater after conventional (primary and secondary) treatment (Semmens and Field 1980).

Summary

The results of the Phoenix studies show that the renovated water from the 23rd Avenue SAT projects meets the public health, agronomic, and aesthetic requirements for unrestricted irrigation, including parks, playgrounds, and vegetable crops that are consumed raw (U.S. Environmental Protection Agency, 1992). The water also meets the standards for lakes with primary contact recreation and for most industrial and other nonpotable uses. Potable use of the renovated water would require additional treatment, for example, reverse osmosis and disinfection. Such treatment, however, would be more effective and economical for renovated water from an SAT system than for effluent from a conventional sewage treatment plant (Bouwer, 1992).

In the Phoenix studies, secondary treated municipal wastewater was used because that was what the treatment plants provided. In general, however, the secondary (biological) treatment step is not necessary because SAT systems can handle relatively large amounts of organic carbon. Thus, where treated municipal wastewater is to be used for a rapid-infiltration system, primary treatment may suffice (Rice and Gilbert, 1978; Lance et al., 1980; Leach et al., 1980; Carlson et al., 1982; Rice and Bouwer, 1984). Some additional clarification or filtration of the primary effluent may, however, be desirable. The higher TOC content of the filtered primary effluent actually may enhance denitrification, removal of recalcitrant organic compounds through secondary utilization and cometabolism (McCarty et al., 1982), and removal of pathogens in the SAT system.

EL PASO, TEXAS, RECHARGE PROJECT

The city of El Paso, located in far west Texas, is an arid environment with an annual rainfall averaging about 20 cm (8 inches). Water supplies are scarce, and the water rights for the Rio Grande are fully appropriated. El Paso obtains about 10 percent of its water supply from the Rio Grande. The remaining 90 percent comes from ground water. Of this, approximately 25 percent of the water supply comes from the Canutillo well field located on the Mesilla Bolson to the west of the Franklin and Organ mountains, and the remaining 65 percent comes from the Hueco Bolson well field located to the east of the mountains (Figure 6.7). (A bolson is a broad and nearly flat mountain-rimmed sediment-filled desert basin with interior drainage.) Juarez, Mexico, which is located directly across the Rio Grande from El Paso, has roughly double the population of El Paso and also depends on ground water from the Hueco Bolson aquifer for its water supply. Because of these two major water users, the freshwater layer of the Hueco Bolson gradually is being depleted and replaced by the more saline ground water that surrounds the fresh ground water resource.

The total water use for El Paso is about 100 mgd. Of this, about 50 percent is returned to sewage plants for treatment. Thus, there is an opportunity to reuse some of this water to recharge the Hueco Bolson aquifer and extend the long-term use of the resource as a potable water supply. The Fred Hervey Water Reclamation Plant provides up to 10 mgd of reclaimed water for ground water recharge. This means that each 10-year period of operation of this facility can extend the resource lifetime of the aquifer by 1 year.

System and Site Description

Type of Recharge

The El Paso recharge project is a 10-mgd direct-injection system that was selected over infiltration basins because of the area's deep water table (about 107 m (350 feet) below the surface). The overall recharge system consists of an advanced wastewater treatmentplant, a pipeline system through the Hueco Bolson, and 10 injection wells. All sewage collected in the northeast of the city is pumped to the treatment plant. Following treatment, the wastewater is pumped to the injection system for injection across the freshwater section of the bolson between existing production wells. After injection, the water travels approximately 1.2 km (.75 mile) through the aquifer to production wells for municipal water supply. In addition, the reclaimed water is available directly for industrial cooling at a nearby plant and some is used to irrigate a city golf course.

Hydrogeologic Conditions

Recharge of the Hueco Bolson aquifer occurs along the foothills of the mountains and plateaus where sediments are coarse grained and permeable. In some places, additional recharge comes from overlying alluvium. The thickness of the Hueco Bolson deposits ranges from 305 to 2,740 m (1,000 to 9,000 ft), with the deepest known freshwater at a depth of about 430 m (1,400 ft).

Before development of the Hueco Bolson as a ground water resource, the

SELECTED ARTIFICIAL RECHARGE PROJECTS 235

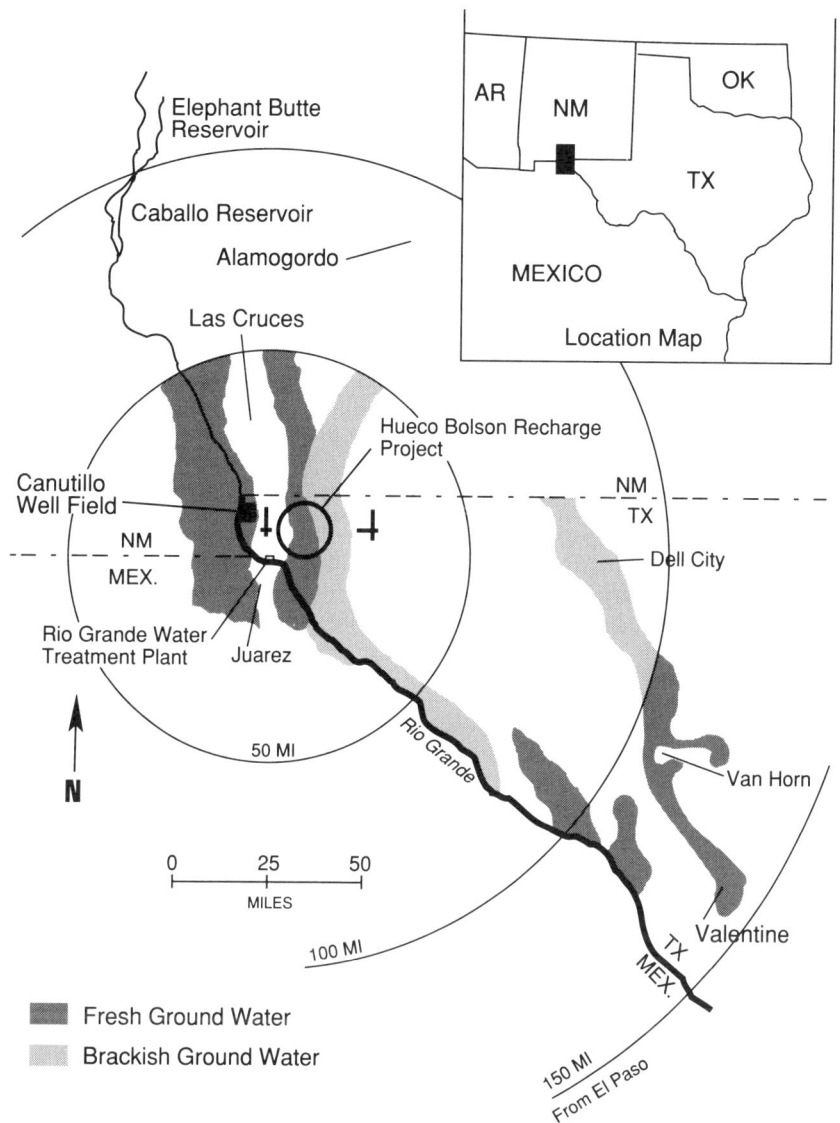

FIGURE 6.7 The Hueco Bolson well field.

floodplain of the Rio Grande served as the discharge zone for the ground water, and the Rio Grande was a gaining stream. The first well supplying ground water to El Paso was drilled around 1892. In 1901 the development of the Mesa well field north of Fort Bliss began, and 44 wells had been completed by 1917. In 1917, development of the Montana well field at a lower elevation near the eastern side of the city began. Subsequently, deep wells were drilled in the Mesa and Montana fields to meet increasing demands. A 1970 water level map shows two cones of depression in the water table east of the Franklin Mountains.

Development of ground water from the bolson deposits near the city has lowered the water level below the water level in the overlying Rio Grande alluvium. Instead of being an area of discharge, the alluvium has become an area of ground water recharge, and the Rio Grande is now a losing stream. Digital modeling suggests that recharge from the Rio Grande alluvium has exceeded the natural recharge rate since the 1950s and 1960s.

Municipal and industrial ground water pumpage has increased steadily since the early 1900s. From 1906 to 1975, about 1.80 million acre-feet was pumped from the Texas part of the northern Hueco Bolson aquifer, whereas from 1925 to 1975 about 570,000 acre-feet was pumped from the Juarez area in Mexico. In 1975, 72,000 acre-feet was pumped from the Texas part of the northern bolson, and about 40,000 acre-feet was pumped from the Juarez area. These pumpages have resulted in a water level decline of 18 to 21 m (60 to 70 ft) in the northern and southern part of El Paso and as much as 29 m (95 ft) in downtown El Paso and Juarez. Irrigation can be a big water user, varying from 10,000 acre-feet per year to as much as 150,000 acre-feet per year during a drought period (Charbeneau, 1982).

The El Paso recharge project involves injection of treated wastewater into the Hueco Bolson, which includes areas in Texas, New Mexico, and Mexico. The injection wells are constructed so that water is recharged from the top of the saturated zone to the point where TDS levels approach 1,000 mg/l. This is generally in the interval from 107 to 290 m (350 to 960 ft). The formation throughout the injection interval is fluvial in nature and contains gravel, silt, and clay lenses. Horizontal hydraulic conductivity of the interval averages about 7.6 m/day (25 ft/day) with the porosity averaging about 20 percent. The transmissivity and specific yield are about 1,225 m^2/day (13,000 ft^2/day and 10 percent. The formation below the injection interval is lacustrine in nature and has a much lower permeability (Knorr and Cliett, 1985).

Source Water

The Fred Hervey Water Reclamation Plant is a wastewater treatment facility designed for recycling water used by the residents in the northeast area of El Paso back to the Hueco Bolson to help meet the water needs of El Paso. The influent characteristics indicate a moderately weak sewage that is primarily domestic in origin and has a minimum industrial component. To avoid the presence of highly toxic compounds in the source water, an extensive public awareness campaign and vigorous enforcement of the industrial waste discharge regulations have been developed and maintained. In addition, there is no chlorination in the wastewater collection system.

The 10-mgd advanced wastewater treatment plant has two 5-mgd parallel process trains which include screening, degritting, primary settling, equalization, a two-stage PACT^R system, lime treatment, sand filtration, ozone disinfection, granular activated carbon filtration, and storage. Generally, the first part of the plant flow scheme concentrates on the removal of organic pollutants, nitrogen, and suspended solids, while the second part provides removal of pathogens, phosphorus, radioactive material, and heavy metals, with additional removal of suspended solids, dissolved organic compounds, and color, taste, and odor problems. In the PACT^R process, the wastewater is aerated so that bacteria can feed on some pollutants while others are adsorbed on powdered activated carbon. This process removes dissolved pollutants including organics and ammonia. Then, in a clarifier, bacteria and powdered carbon are separated from the water, and methanol is added for denitrification. The lime process at a pH level of 11.1 provides high viral and heavy metal removal performance. This process serves as the first step in the disinfection process. Ozone is used as the final disinfecting process because of its advantages over chlorine in disinfection performance and to avoid the formation of halogen, a disinfection by-product (DBP). A very low (0.25 mg/l) chlorine dose is provided ahead of storage to prevent buildup of film or growths in the clear well during the 8-hour storage periods. Granular activated carbon (GAC) filtering is incorporated in the process train as a polishing process in the removal of residual organic compounds (Knorr and Cliett, 1985).

Operational History

The basic objective of the El Paso recharge project was to increase potable water supplies with the lowest practicable risk. Two basic criteria for design were to provide (1) maximum recovery of recharged water to minimize the costs and (2) adequate aquifer residence time to provide the opportunity for additional purification in the aquifer. The El Paso Water Utilities ground water recharge project has been operating since 1985 the recharge to Hueco Bolson aquifer. Ground water movement studies indicate that a downgradient spacing of 1,200 feet between injection and production wells would provide a residence time of 2 years. The design aquifer residence time of 2 years was chosen based on viral inactivation periods, which are on the order of months. The actual spacing in the field has the injection wells located approximately 1.2 km (.75 mile) upgradient from the existing production wells. This spacing should give residence times of at least 6 years. In addition, where future production wells are drilled to fill in a

5 well per section pattern, injection well to production well spacing will still exceed 370 km (1,200 ft). Actual monitoring of breakthrough at observation and recovery wells has proved to be difficult because of the similarity between recharge and formation water qualities. Subsequent field monitoring suggests that the minimum residence time is closer to 7 years (P. Buszka, personal communication, 1993). The produced water is chlorinated before reuse, with no other treatment. To date, there have been no operational difficulties (J. Balliew, personal communication, 1993).

Available Performance Data

Quantity

The injection rate of 700 gpm for the wells is one-half to two-thirds of the capacity of production wells in the area, allowing for a decrease in well efficiency with time. The wells are periodically pumped for back-flushing and cleaning. There have been no long-term performance problems (J. Balliew, personal communication, 1993).

Quality

The quality of the water produced by the Fred Hervey Water Reclamation Plant is tested continually. Table 6.3 shows that the water produced is comparable to the water currently in the Hueco Bolson aquifer and that it meets state and federal regulatory levels for safe drinking water.

An ongoing U.S. Geological Survey study is focusing on the fate of disinfection by-products and on tracing injection waters. Results from monitoring have suggested that tribromomethane concentrations decrease with distance from injection wells. Associated with a decrease in tribromomethane concentrations is an increase in dibromomethane concentrations, although the mechanisms responsible for this have not been identified (P. Buszka, personal communication, 1993).

Economic and Institutional Considerations

A breakdown of construction costs for the treatment plant and recharge system is shown in Table 6.4. Although this table is based on a $27 million bid cost, the final cost was $33 million.

Project managers selected the northernmost well fields serving El Paso for recharge because they offered the greater potential benefits to the city than the other well fields in the aquifer. The city of El Paso's Public Service Board owns most of the water in this locality and would receive the most benefit from an artificial recharge program. Further, the location of the recharge and production

TABLE 6.3 Water Quality at Hueco Bolson Aquifer and Fred Hervey Reclamation Plant

	Ground Water from Hueco Bolson	Reclaimed Water from Hervey Plant
Bicarbonate	220.0	220.0
Chloride	105.0	171.0
Fluoride	1.09	0.90
Nitrate	7.10	1.60
Phosphate	0.03	3.07
Silica	25.0	34.0
Sulfate	67.0	85.0
Barium	0.03	0.014
Calcium	51.0	61.0
Iron	0.15	<0.10
Magnesium	14.0	4.42
Manganese	0.05	<0.05
Potassium	4.80	15.50
Sodium	83.0	164.0
Hardness (calcium carbonate)	184.0	167.0
pH	7.92	7.60
Total dissolved solids	598.0	670.0
Turbidity (NTUs)	0.44	0.14

Note: All units are mg/l except pH and turbidity. NTU is nephelometric turbidity units.

Source: El Paso Water Utilities, undated.

wells suggests that virtually all of the recharged water will be recovered by the city because gradients are generally toward the city's production wells.

The discharge permit from the Texas Water Commission requires the monitoring of 23 variables, with 30-day average values to be used on most variables. Monitoring frequency also is specified. The permit limits are the same as the drinking water standard. The permit also requires less than 10 mg/l nitrate (N) and less than 5 NTU turbidity in each 8 hour batch of water.

Summary

El Paso is in an arid environment with a limited water supply and a water resource that is being depleted. Water management is a necessity. Ground water recharge operations are extending the lifetime of the Hueco Bolson aquifer. Water reclaimed from sewage is being injected with the intent of recovery for potable reuse with only chlorination. Monitoring suggests that injection waters are only now reaching production wells after seven years of production. No

TABLE 6.4 Construction Costs for the El Paso Treatment Plant and the Recharge System

Treatment Plan	Construction Costs ($1,000)	Amortized Capital ($/1,000 gal)	Estimated O & M ($/1,000 gal)	Total Cost ($/1,000 gal)
Screen and degrit	413	0.01	0.01	0.02
Primary treatment	2,419	0.08	−0.01[a]	0.07
Equalization	283	0.001	0.01	0.011
PACT System	13,637	0.48	0.41	0.88
Lime and recarbonation	3,052	0.11	0.35	0.46
Sand filtration	668	0.02	0.003	0.023
Ozonation	1,016	0.035	0.03	0.065
GAC filtration	1,752	0.061	0.03	0.09
Storage	2,676	0.09	0.002	0.092
Chlorination	193	0.007	0.003	0.01
Injections Wells				
Pipelines	2,300	0.07	0.03	0.10
Wells	2,300	0.05	0.009	0.059
Total cost				$1.88

[a]Negative due to gas production.

Source: Knorr and Cliett, 1985.

significant problems have been identified during the relatively short operational history of this project.

LONG ISLAND, NEW YORK, RECHARGE BASINS

Urbanization of Nassau and Suffolk counties on Long Island, with the attendant construction of highways, houses, shopping centers, industrial areas, streets, and sidewalks in previously undeveloped or agricultural areas, has caused a twofold water management problem: (1) disposal of stormwater runoff from impervious areas and (2) reduction in land-surface area available for infiltration of precipitation to naturally recharge the ground water reservoir. To obviate the need for costly trunk storm sewers to convey runoff to streams or coastal waters and to minimize the loss of recharge, shallow stormwater collection basins have been built to contain stormwater runoff and allow it to infiltrate to the water table. These stormwater infiltration basins have been used since 1935. Their number increased as development progressed on Long Island—from 14 basins in 1950, to more than 700 in 1960, to more than 2,100 in 1969, to more than 3,000 by 1986 (Seaburn and Aronson, 1974; Ku and Simmons, 1986). The basins are located mainly in eastern Nassau County and western Suffolk County in the inner part of Long Island.

In addition to facilitating the disposal of stormwater runoff in developed areas, the basins provide recharge to the ground water reservoir, which is the sole source of water for the 2.7 million inhabitants of Nassau and Suffolk counties. More than 10 percent of the 3,070 square-km (1,200 square-mile area) in Nassau and Suffolk counties drains to recharge basins (Ku and Simmons, 1986). The collection and routing of stormwater from impervious areas to these basins have countered the loss of land-surface area available for infiltration of precipitation. In areas drained to the basins, ground water recharge from precipitation is probably equal to or slightly greater than recharge under predevelopment conditions (Seaburn and Aronson, 1974). Recharge provided by the stormwater basins partially replenishes the ground water withdrawn for use by Long Island residents and thereby helps retard seawater intrusion into the aquifers and the drying up of streams.

System and Site Description

Type of Recharge

The Long Island recharge basins are a system of unlined pits of various sizes and shapes excavated in moderately to highly permeable surficial sand and gravel deposits. They range in size from 0.1 to 30 acres and average between 1 and 2 acres. Most extend 3.1 to 4.6 m (10 to 15 ft) below land surface, but some are as deep as 12 m (40 ft) (Seaburn and Aronson, 1974). For the most part, the basins are built in areas where the water table is sufficiently deep to remain below the floor of the basin most of the time.

Most of the basins are constructed with overflow structures that are not more than 3.1 m (10 ft) above the floor of the basin. The required capacity of the basin below the overflow altitude is estimated by multiplying the volume of water equivalent to 13 cm (5 inches) of rainfall on the total area drained to the basin by a factor ranging from 30 to 100 percent based on drainage area conditions such as land slope and percentage of paved area. A factor of 30 percent is used in most residential areas, whereas for industrial areas with a higher proportion of impervious surface the factor is as much as 100 percent. In calculating the design capacity, infiltration into the floor and sides of the basin during inflow is not considered, thereby providing a safety factor.

Various construction features are used to ensure operational efficiency. These include multilevel basin floor, retention basins, wells, and scarification of basin floors. The lower levels of a basin floor act as a settling area for inflowing sediment and trash and allow higher infiltration rates in the higher-level floor areas receiving relatively sediment-free water as overflow. Retention basins serve the same purpose as basins with integral settling areas except that they are separate basins connected by pipes or channels to adjacent or nearby basins. Where deposits of low hydraulic conductivity immediately underlie the basin floor, wells often are installed beneath the basin floor to provide access to deeper-lying more permeable strata. Scarification of basin floors is also used to expose more permeable underlying deposits.

Hydrogeologic Conditions

The surficial deposits into which the recharge basins are excavated consist mainly of unconsolidated outwash deposits and ground and terminal moraine deposits of the Wisconsin Glaciation. The outwash deposits are well-sorted sand and gravel of moderate to high permeability. The morainal deposits consist of poorly sorted till of moderate to low permeability. The surficial deposits overlie unconsolidated interbedded sand, gravel, silt, and clay of Cretaceous age. The Cretaceous deposits have high to low permeabilities, depending on lithology. The water supply for Nassau and Suffolk counties is obtained totally from various aquifers composed of both the glacial and the cretaceous sediments.

Source Water

Water recharged by the Long Island basins is stormwater runoff from residential, industrial, and commercial areas and from highways. As such, it is highly variable in both quantity and quality. Quantity varies depending on the intensity and duration of the storm as well as the percentage of impervious area within the drainage basin. Quality varies depending on the land use activities within the drainage area, length of time of the runoff event, and time since the last runoff event. The stormwater runoff is piped directly to the recharge basin and receives no treatment except for the settling of sediment.

Operational History

Most of the basins in which stormwater is impounded are commonly dry within a day or so after a storm, but some hold water perennially (Aronson and Seaburn, 1974). Of the basins existing on Long Island in 1969, 9 percent were found to contain water more than 5 days after a 2.5 cm (1-inch) rainfall on the drainage area (Aronson and Seaburn, 1974). Containment of water in these basins over prolonged periods occurs for one or more of these reasons: (1) they intersect either the regional or a perched water table, (2) they are excavated in materials of low permeability, or (3) sediment and debris accumulating on the basin floor reduce infiltration rates.

A study of water-containing basins revealed that those that drain commercial and industrial areas had the highest percentage, 28 percent, of clogging caused by sediment and debris. Only 7 percent of residential area basins and 9 percent of highway basins were clogged and thus contained water. The high percentage of water-containing commercial and industrial basins is probably largely related to the large influxes of asphalt, grease, oil, tar, and rubber particles in storm runoff from adjacent parking fields (Aronson and Seaburn, 1974). Although basins that drain highways might be expected to receive similar materials, the small percentage of highway basins that contain water probably reflects a combination of the more regular maintenance of these basins and the use of special structures incorporated into the basin floors.

Performance Data

Quantity and Hydraulics

The 2,124 recharge basins operating on Long Island in 1969 recharged an estimated annual total of 68,100 acre-feet of water (Seaburn and Aronson, 1974). Infiltration rates were studied at three typical basins in Westbury, Syosset, and Deer Park. At Westbury infiltration rates computed from data collected during 63 storms ranged from 9.1 to 52 cm/hr (0.3 to 1.7 ft/hour) and averaged 27 cm/hour (0.9 ft/hour). At Syosset, rates during 22 storms ranged from 9.1 to 55 cm/hour (0.3 to 1.8 ft/hour) and averaged 24 cm/hour (0.8 ft/hour). At Deer Park, rates during 24 storms ranged from 3.0 to 15 cm/hour (0.1 to 0.5 ft/hour) and averaged 6.0 cm/hour (0.2 ft/hour) (Seaburn and Aronson, 1974). These rates were observed under a wide range of meteorological conditions.

Quality

Five recharge basins representing different land use areas were monitored during 46 storms to determine quality of stormwater runoff and precipitation and quality of ground water immediately beneath the basins 1 or 2 days after the storm. Samples were analyzed to identify standard inorganic constituents, heavy metals, organic compounds, and bacteria. Conclusions from this study included (Ku and Simmons, 1986):

- Most of the load of heavy metals in the stormwater was removed during infiltration through the unsaturated zone, but nitrogen and chloride were not removed.
- The median number of indicator bacteria in stormwater ranged from 10^8 to 10^{10} MPN (Most Probable Number) per 100 milliliter. Virtually no bacteria were detected in ground water beneath the recharge basins, indicating complete removal during percolation of stormwater through the unsaturated zone.
- Concentrations of pesticides in basin-bottom soils generally were much higher than those in stormwater, suggesting that pesticides are probably sorbed or filtered out in the soil layer.
- Use of recharge basins on Long Island to dispose of stormwater runoff

and to recharge ground water does not appear to have significant adverse effects caused by chemical and microbiological constituents.

Economic and Institutional Considerations

There is no pre-recharge treatment of stormwater other than the settling of solids that is accomplished by the initial construction of settling areas on the basin floors or by separate retention basins. Thus the cost of the recharge operations is limited to land acquisition, basin construction, and basin maintenance. Basin maintenance consists mainly of collecting and removing bulk debris and cutting and removing grass on the basin floor. Scarification of basin floors is done as required to break or loosen material on the basin floor or to remove a thin layer of clogging material.

The water recharged via these basins is not directly recovered. Rather, it becomes mixed with the water naturally recharging the ground water reservoir and is recovered by numerous public and private supply wells distributed throughout Nassau and Suffolk counties.

Most of the basins on Long Island are owned and maintained by local or state governmental agencies. Developers are required to construct storm sewers and recharge basins of adequate size for their project. On completion, recharge basin ownership and maintenance become the responsibility of the local government.

Summary

Ground water recharge of stormwater has proved to be a viable means of locally disposing of storm runoff in a rapidly urbanizing part of Long Island and, at the same time, countering the reduction of natural recharge to the ground water system caused by the increase in impervious areas. The Long Island aquifer system has been designated by EPA as the "sole-source aquifer" for water supply in Nassau and Suffolk counties, a 3,070 square km (1,200 square mile) area having a population of 2.7 million. Studies have not shown any significant adverse impact on ground water quality, even after several decades of operation for some of the early basins. The stormwater receives no treatment other than the settling of sediment before it infiltrates, which for most basins occurs within hours to a few days following the storm event. Studies also have shown that in areas served by drainage basins, ground water recharge is equal to or slightly exceeds that occurring under predevelopment conditions. The storage and flow accretions to the aquifer system realized through the recharge of stormwater appear to far outweigh any potentially detrimental water quality impact.

ORLANDO, FLORIDA, STORMWATER DRAINAGE WELLS

Drainage wells to dispose of excess surface water in the city of Orlando and surrounding areas have been in operation since 1904 (Kimrey and Fayard, 1984). The drainage wells alleviate flooding and control lake levels in this area of closed drainage basin lakes and low topographic relief. Although drainage wells are used in other parts of Florida also, the Orlando area has the highest concentration of wells (Kimrey and Fayard, 1984). These wells are used to facilitate drainage in this internally drained karst environment and also to recharge the Floridan aquifer system, the primary source of water for the Orlando area and most of Florida.

System and Site Description

Type of Recharge

In 1990, there were about 310 drainage wells within the greater Orlando area (Figure 6.8), an area of about 230 square km (90 square miles) (Bradner, 1991). These wells inject surface water by gravity into the Upper Floridan aquifer. More than half of the wells are 30 cm (12 inches) or more in diameter, but they range from 10 to 61 cm (4 to 24 inches) in diameter. The drainage wells range in depth from about 37 to 320 m (120 to 1,050 feet); median depth is about 120 m (400 ft). They are cased to or near the top of the Upper Floridan aquifer and then finished as an open hole in the aquifer.

Hydrogeologic Conditions

The Orlando area is underlain by about 15 m (50 ft) of sand and silt that constitute the surficial aquifer. The surficial aquifer is, in turn, underlain by about 46 m (150 ft) of sandy clay, silt, and shell that constitute the intermediate confining unit. Beneath the intermediate confining unit lies the Floridan aquifer system, which is made up of about 460 m (1,500 ft) of limestone and dolomite. The Floridan aquifer system has been subdivided into three units—the Upper Floridan aquifer (91 to 120 km (300 to 400 ft) thick), the middle semiconfining unit of less permeable limestone (91 to 180 m (300 to 600 ft) thick), and the Lower Floridan aquifer (120 to 180 m (400 to 600 ft) thick). These aquifers have a lot of secondary porosity (karstic) and are very permeable; within the Orlando area, transmissivity of the Upper Floridan aquifer ranges from 4,700 to 37,000 square m per day (50,000 to 400,000 square feet per day), and transmissivity for the Lower Floridan aquifer ranges from 9,200 to 57,000 square m per day (100,000 to 600,000 square feet per day) (Tibbals, 1990). Water supply for the Orlando area is obtained mainly from the Lower Floridan aquifer, but some is pumped from the Upper Floridan aquifer as well. Virtually all of the drainage

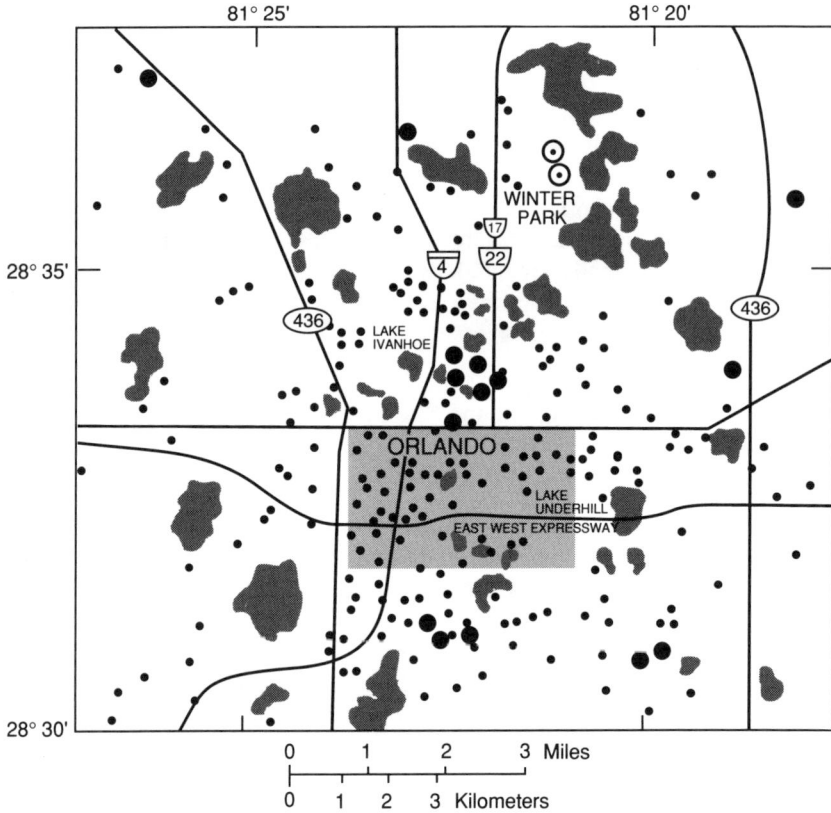

FIGURE 6.8 Location of drainage wells and public-water supply wells within the Orlando area.

Source: Bradner, 1991.

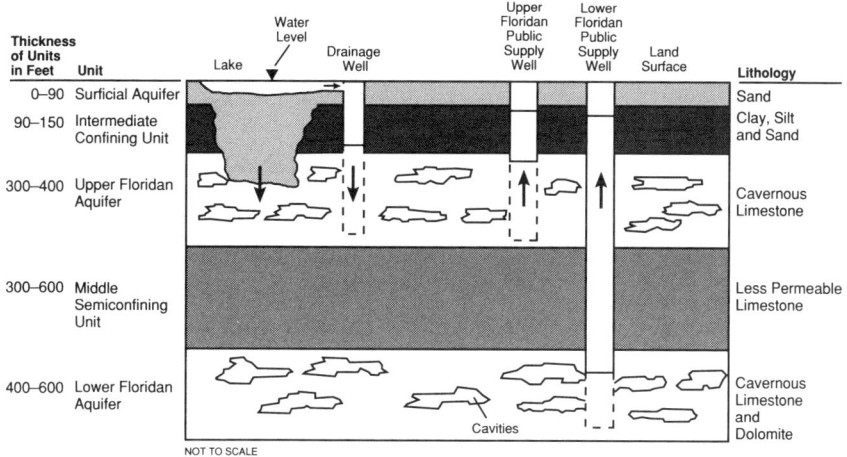

FIGURE 6.9 Generalized hydrogeologic section in the Orlando area Bradner, 1991.

wells are completed in the Upper Floridan aquifer. Figure 6.9 shows a generalized hydrogeologic section beneath the Orlando area.

The water table in the surficial aquifer is higher in altitude than is the potentiometric surface of the Upper Floridan aquifer. However, the low-permeability materials in the intermediate confining unit impede the downward flow of ground water from the surficial aquifer to the Upper Floridan aquifer. The drainage wells "short circuit" the intermediate confining unit and provide for direct input of surface water to the Upper Floridan aquifer.

Source Water

About 50 percent of the 310 drainage wells receive stormwater runoff directly from streets or other impervious areas, 45 percent receive lake or wetland overflow, and 5 percent receive air-conditioning return water or are unused at present but have received industrial effluent or sewage in the past. The water receives no treatment prior to injection. Grates, either directly on the wellhead or on the intake chamber leading to the well, screen coarse debris carried by the water. Water flowing directly from storm sewers is available intermittently during storm events. Many lake-overflow wells receive water continuously.

Operational History

The cavity riddled, highly permeable nature of the limestone of the Upper Floridan Aquifer allows the drainage wells to function more or less unattended. However, some wells are reported to have been completely filled by debris and sand, and many wells must be cleaned periodically to maintain their effectiveness.

The potentiometric surface of the Upper Floridan aquifer normally is well below land surface throughout most of the area, so sufficient head is available for the drainage wells to accept large quantities of water under gravity feed. However, during very wet years, such as 1959 and 1960, heads in the Upper Floridan aquifer increased to the point that the capacity of the drainage wells to recharge surface water to the aquifer decreased. In fact, some drainage wells actually flowed, and pressure injection pumps had to be installed to continue to use the wells for recharge until the potentiometric surface once again declined to below land surface (Kimrey, 1978).

Performance Data

Quantity

Limited quantitative data are available on acceptance rates of drainage wells, but the range is reported as a few hundred to several thousand gallons per minute. An acceptance rate as high as 9,500 gpm was reported for a drainage well in west Orlando (Stringfield, 1933). Collectively, the 310 drainage wells in the Orlando area are reported to recharge an estimated 23 mgd of surface water to the Upper Floridan aquifer, but this is probably a low estimate (Bradner, 1991). A lake overflow well gaged from November 1987 through December 1988 averaged an inflow rate of 2.1 mgd, whereas a stormwater well reportedly accepted an average of 9,000 gpm during 1988 (Bradner, 1991).

Quality

Quality of inflow to the drainage wells has been summarized by Bradner (1991). According to that review, stormwater runoff contains high concentrations of total organic carbon, organic nitrogen, iron, lead, sulfate, and zinc, while concentrations of most anions and cations are lower in stormwater runoff than in water from the Upper Floridan aquifer (Wanielista et al., 1981; German, 1989). Concentrations of total organic carbon can range from 18 to 284 mg/l in runoff to Lake Eola in downtown Orlando (Wanielista et al., 1981). German (1989) found that inflow to the drainage wells frequently had detectable concentrations of many pesticides, with diazinon being detected in 77 percent of the samples collected and malathion being detected in 50 percent of the samples.

Studies of loads of nutrients and organic compounds entering the Upper Floridan aquifer at nine drainage well sites indicate that approximately 45,000 kg (100,000 pounds) of total nitrogen enters the Upper Floridan aquifer each year through drainage wells in central Florida (German, 1989). Studies of stormwater runoff also have reported sporadic detections of phthalates, compounds widely used in the plastics industries, and polycyclic aromatic hydrocarbons, such as fluoranthene, pyrene, anthracene, chrysene, and benzo-a-pyrene, commonly associated with petroleum products (Wanielista et al., 1981; German, 1989).

The quality of water in the Upper Floridan aquifer in the Orlando area also has been studied. Schiner and German (1983) concluded that drainage wells and upper-producing-zone supply wells yielded water very similar in chemical characteristics, particularly major dissolved constituents. Water in the upper producing zone of the Floridan aquifer is primarily a calcium and magnesium-bicarbonate type. Bicarbonate generally accounts for more than 75 percent of the ions, and calcium and magnesium account for more than 85 percent of the cations. But in several supply wells, and several drainage wells, more than 25 percent of the anions consisted of sulfate plus chloride, and more than 15 percent of the cations consisted of sodium plus potassium. Water from the lower producing zone (also a calcium and magnesium-bicarbonate type water) was more consistent within its chemical type. This consistency may be because most samples from the lower producing zone were clustered in a small part of the study area or it may be because the zone is deeper and more isolated from surface influences.

The study also noted that water from drainage wells generally has slightly higher concentrations of most constituents than water from supply wells. The primary differences in water quality between drainage wells and supply wells were for total nitrogen, total phosphorus, total recoverable iron, and total coliform. The comparisons are shown in Table 6.5.

For some supply and drainage wells, color, hydrogen sulfide, iron, and manganese in these studies exceeded the National Secondary Drinking Water Regulations, with the frequency of exceedance greater for drainage wells than for supply wells. Concentrations of metals and pesticides did not exceed the limit specified in Florida standards for potable ground water. Pesticide did not appear to be present in significant amounts.

Overall, the quality of water from the group of supply wells in the Orlando area is about the same as the quality of water from wells in adjacent areas where

TABLE 6.5 Differences in Water Quality for Two Types of Wells in Orlando, Florida

	Drainage Wells	Supply Wells
Total nitrogen	1.0 mg/l	0.29 mg/l
Total phosphorus	0.23 mg/l	0.07 mg/l
Total recoverable iron	660 mg/l	60 µg/l
Total coliform	39 per 100 mg/l	0 per 100 ml

Source: Schiner and German, 1983.

no drainage wells exist. Water quality for drainage wells that receive street runoff was about the same as water quality for drainage wells that receive lake overflow, except for bacteria colony counts. Bacteria counts were considerably lower in wells that receive lake overflow than in those that receive direct street runoff.

Results of the Schiner and German (1983) study indicate that drainage well recharge has not caused widespread contamination of the Floridan aquifer. Bacterial contamination found in some drainage wells appears highly localized, and water from drainage wells would generally be acceptable for public supply use as long as bacteria are not present. Another study (Bradner, 1991) of 11 supply wells in urban Orlando, where the highest density of drainage wells exists, found calcium, potassium, sodium, chloride, and ammonia in significantly higher concentrations than is samples from hydrogeologically similar areas elsewhere. Significant differences in other constituents were not indicated.

Hydraulics

Specific capacity data under pumping conditions are available for 21 drainage wells. At pumping rates ranging from 240 to 460 gpm, the wells reportedly had specific capacities that ranged from 27 to 1,900 gpm/feet, with the median being 310 gpm/feet. The 23 mgd of recharge from the wells in the Orlando area has created a mound in the potentiometric surface of the Upper Floridan aquifer of 1.2 m (4 ft) (Tibbals, 1990).

Economic and Institutional Considerations

The stormwater used in the Orlando drainage wells does not receive any pre-recharge treatment except for that provided by detention in lakes for wells receiving lake overflow, and so pre-recharge treatment costs are limited. In addition, operating costs are low because the system operates without attention except for the infrequent need to clean out accumulated debris and sediment from the wells in order to maintain their efficiency.

Most drainage wells are owned by municipalities or the Florida Department of Transportation. They are regulated as Class V injection wells under the Safe Drinking Water Act. No new drainage wells are currently being permitted.

Summary

Drainage wells are the most economical way of disposing of stormwater in the internally drained karst environment of the Orlando area. The drainage wells emplace, by gravity injection, 23 mgd of recharge to the Upper Floridan aquifer, which helps to balance the 51 mgd of municipal ground-water pumpage in the Orlando area. Some drainage wells accept urban stormwater runoff directly from street drains, whereas others accept overflow from lakes into which stormwater has drained. Therefore, they introduce contaminants directly to one of the aquifers used for public supply in the area. Although some water quality effects have been noted, widespread contamination has not occurred, even though usage of drainage wells began in 1904. However, at least one plume of contaminated water in the vicinity of a drainage well is known. Spills of chemicals and/or fuels caused by accidents along transportation routes are possible and pose a risk of introducing highly concentrated contaminants into the aquifer.

Alternate means of stormwater disposal in this karst area would require extensive trunk sewers and pumping at considerable cost. Moreover, loss of the recharge provided by the drainage wells would result in a reduction of head in the Floridan aquifer system and the possibility of vertical encroachment of deeper-lying saltwater into supply wells. The current level of risk of severely contaminating the potable source aquifer is accepted, but no new drainage wells are being permitted.

DAN REGION WASTEWATER RECLAMATION PROJECT METROPOLITAN TEL AVIV, ISRAEL

Where water is scarce, municipal wastewater can serve as an unconventional source of supply that can be integrated into the regional water supply system. In Israel, the increased demands for high-quality water and the shortage of natural water sources have resulted in the development of strategies to improve the quality of secondary effluent to make it suitable for nonpotable uses, especially unrestricted agricultural reuse. The best example of this approach is the Dan Region Wastewater Reclamation Project,* which provides for the collection, treatment, recharge, and reuse of the wastewater from the largest metropolitan area of the country, including Tel Aviv–Jaffa and several other neighboring municipalities. The project serves a total population of about 1.3 million with an average municipal wastewater flow of 72 million gallons per day.

The recharge-recovery method developed and practiced in the Dan Region project relies on the soil-aquifer treatment (SAT) concept. Partially treated effluent percolates through the unsaturated soil zone (fine sand) until it reaches the ground water. It moves radially in the aquifer until it reaches recovery wells designed to pump the recharge water for supply (Figure 6.10). Depths for ground water range from 15 to 45 m (49 to 150 ft) for the various sites. Distances between recharge basins and recovery wells range from 320 to 1,500 m (1050 to 4900 ft). If the recovery wells are adequately spaced, the recharge and recovery facilities can be operated to confine the recharged effluent between the recharge

---

*The committee would like to thank M. Michail and A. Kanarek of Mekorot Water Co., Israel, for their efforts in compiling the information in this section.

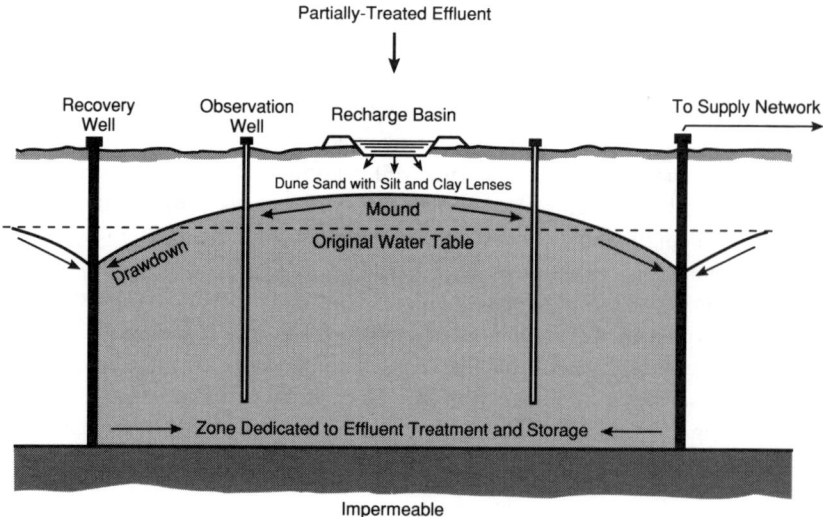

FIGURE 6.10 SAT system in the Dan Region project, Israel.

areas and the recovery wells. This underground subbasin is dedicated to the treatment and storage of effluent and represents only a small percentage of the regional aquifer. The recharge water, which can be traced and monitored by means of observation wells, is usually of high quality. It is generally appropriate for industrial uses, unrestricted agricultural uses (including irrigation of vegetables to be eaten raw and livestock watering), nonpotable municipal uses, and recreational uses. Accidental drinking of recharge water does not present a significant health hazard because of its high microbiological quality.

Recharge is done with spreading basins to take advantage of the purification capacity of both the unsaturated zone and the aquifer. Operation of the recharge basins is intermittent; flooding periods are alternated with adequate drying periods to maintain high infiltration rates and to allow oxygen penetration into the soil to enhance the purification capacity of the system. Although most of the purification takes place during vertical flow through the upper soil layer and the whole unsaturated zone, additional purification (mainly breakdown of slowly biodegradable organics) is gained during horizontal flow in the aquifer, and aerobic properties that were lost in sections where anoxic conditions prevail below the recharge basins are regained.

System and Site Description

Stage one of the Dan Region project has been in operation since 1977. For the first 2 years, the wastewater underwent biological treatment in oxidation ponds with recirculation and chemical treatment by the high lime-magnesium process; the water then moved to polishing ponds for partial free ammonia stripping and natural recarbonation. The combination of oxidation ponds and lime treatment is roughly equivalent to secondary treatment, although the effluent quality is different in some respects. In October 1989, chemical treatment was discontinued and the oxidation pond effluent was conveyed to the polishing ponds.

Stage two of the project has been in operation since 1978. About 60 percent of this wastewater is conveyed to a mechanical-biological treatment plant, where it undergoes primary treatment and secondary treatment by activated sludge with nitrification-denitrification. The remainder goes through oxidation and polishing ponds, parallel to the mechanical-biological treatment plant. The long detention times in the polishing ponds produce a high-quality effluent with relatively low algae content.

The recharge sites are located in areas of rolling sand dunes near the Mediterranean coast underlain by a calcareous sandstone aquifer—one of the three main potable water supplies of the country. The climate of the zone is typically Mediterranean. Summers are warm and dry, and winters are mild with rainy spells. The average annual precipitation is 500 to 600 mm (20 to 24 in). The average temperatures usually range between 20 and 30°C in summer and between 10 and 20°C in winter.

The recharge operation in the second stage is carried out at two sites located south of the treatment plant. One recharge site consists of four basins covering a net area of about 59 acres; each basin is divided into four subbasins. The depth of the unsaturated zone below the recharge basins varies between 27 and 36 m (89 to 120 ft). A ring of recovery wells spaced 300 to 400 m apart surrounds the recharge areas on the northern, western, and southern sides; they are located between 320 and 1,100 m (1,050 to 3,600 ft) from the nearest recharge basin. At some locations, two separate wells are drilled to different subaquifers or to different layers of the same subaquifer. A monitoring network of 20 observation wells was established between the recharge basins and the recovery wells; they are located between 20 and 570 m (65 to 1,870 ft) from the recharge basins.

The second recharge site consists of three basins covering a net area of about 44 acres; each basin is divided into three subbasins. The depth of the unsaturated zone below the recharge basins varies between 40 and 43 m (131 to 141 ft), and it is similar to the first recharge site. A ring of recovery wells surrounds the recharge areas located between 350 and 1,500 m (1,150 to 4,900 ft) from the nearest recharge basins. A monitoring network of 12 observation wells was established between the recharge basins and the recovery wells; they are located between 20 and 300 m (65 and 980 ft) from the recharge basins.

The spreading basins are flooded intermittently to maintain high infiltration rates and to enhance effluent purification during percolation. The water depths in the basins are generally below 0.6 m (2.0 ft). A short recharge cycle is employed, usually consisting of 1 day flooding and 2 to 3 days drying, to ensure that aerobic conditions predominate in the unsaturated zone and in the aquifer.

The clogging layer on the basin bottom may reach a thickness of about 1.5 cm (0.6 in). The clogging layer is "scratched" about once a week or every 2 weeks, especially in the winter when drying is slow, and "shaved off" completely every 1 or 2 months to restore infiltration capacity. There is no accumulation of clogging material deeper in the soil.

Soil-Aquifer Treatment

The major purification processes occurring in the soil-aquifer system are slow-sand filtration, chemical precipitation, adsorption, ion exchange, biological degradation, nitrification, denitrification, and disinfection. To illustrate the purification effect of soil-aquifer treatment (SAT), quality data were evaluated for the mechanical-biological plant effluent (RE) before recharge and SAT and for the reclaimed water (RW) after SAT (Tables 6.6., 6.7, 6.8).

TABLE 6.6 SAT Performance: Basic Wastewater Parameters (averages for 1990)

	Units	Before SAT (RE)	After SAT (RW)	Percentage Removal
Suspended solids	mg/l	17	0	100
Biochemical oxygen demand (BOD)	mg/l	19.9	< 0.5	> 98
BOD filtered	mg/l	3.1	< 0.5	> 84
Chemical oxygen demand (COD)	mg/l	69	12.5	82
COD filtered	mg/l	46	12.5	73
Total organic carbon	mg/l	20	3.3	84
Dissolved organic carbon	mg/l	13	3.3	75
UV 254 absorbance	cm$^{-1} \times 10^3$	298	64	79
$KMnO_4$ as O_2	mg/l	14.1	2.3	84
$KMnO_4$ filtered as O_2	mg/l	12.6	2.3	82
Detergents	mg/l	0.5	0.078	84
Phenols	/Jgll	8	< 2	> 75
Ammonia, as N	mg/l	7.56	< 0.05	99
Kjeldahl nitrogen	mg/l	11.5	0.56	95
Kjeldahl nitrogen filtered	mg/l	10.2	0.56	95
Nitrate	mg/l	2.97	7.17	
Nitrite	mg/l	1.24	0.10	92
Nitrogen	mg/l	15.7	7.83	50
Nitrogen filtered	mg/l	14.4	7.83	46
Phosphorus calcium	mg/l	3.4	0.02	99
Alkalinity, as calcium carbonate	mg/l	306	300	-
pH	-	7.7	7.9	-

TABLE 6.7 SAT Performance: Bacterical and Viral Quality (averages for 1990)

	Units	Before SAT (RE)	After SAT (RW)
Total bacteria	No./ml	110,000	288
Coliforms	MPN/100 ml	1,100,000	0
E. coli	MPN/100 ml	130,000	0
Streptococcus faecalis	MPN/100 ml	29,000	0
Enteroviruses	PFU/200 l	2	0

Note: PFU = plaque-forming units; MPN = Most Probable Number.

Basic Wastewater Parameters

The relatively high removal efficiency obtained for a variety of parameters confirms that SAT is an integral part of the municipal wastewater treatment process in the Dan Region project. The removal of suspended solids (mostly organics) and of biochemical oxygen demand was virtually complete. The average total and filtered chemical oxygen demand were reduced from 69 and 46 mg/l respectively, to 12.5 mg/l in each case. Average total organic carbon and dissolved organic carbon were reduced from 20 and 13 mg/l respectively, to 3.3 mg/l. Ultraviolet absorbance was reduced significantly. The concentration of detergents was reduced from 0.5 to 0.08 mg/l, and that of phenols from 8 to less than 2 µg/l. Because of the efficient and reliable removal of organics, the soil-aquifer system can be regarded as a biological treatment unit.

Total and filtered nitrogen were reduced from 15.7 and 14.4 mg/l, respectively, to 7.8 mg/l. Ammonia was reduced from 7.6 mg/l as nitrogen to less than 0.95 mg/l. While most nitrogen in the recharge water is found in the unoxidized forms of ammonia and organic nitrogen, the residual nitrogen in the well water consists essentially of nitrates. Thus complete nitrification and partial denitrification occur in the soil-aquifer system. Phosphorous removal efficiencies were 25 percent in the oxidation and polishing ponds and 74 percent in the mechanical-biological treatment plant. The remaining phosphorous was removed efficiently by SAT from 3.4 mg/l in the recharge water to 0.02 mg/l in the recovered water, a concentration is similar to that in the natural ground water. Coliform bacteria, E. coli, S. faecalis, and enteroviruses were not detected in the recovered water.

Irrigation-Water Quality Parameters

The salinity of the recovered water is acceptable for unrestricted irrigation of all crops. The sodium adsorption ratio (SAR) in the recovered water is similar to that of the recharge water (about 4.6), which is acceptable for unrestricted

TABLE 6.8 SAT Performance: Irrigation Water Quality Parameters (averages for 1990)

	Units	Before SAT (RE)	After SAT (RW)	Percentage Removal	Tolerance for water used continuously on all soils[a]
Salinity and Sodium Hazard					
Chloride	mg/l	293	276		
Dissolved Solids	mg/l	1,033	1,033		
Electrical conductivity	μmhos/cm	1,642	1,597		
Sodium	mg/l	194	203		
Potassium	mg/l	23	16		
Sodium absorption ratio		4.6	407	30	
Trace Elements					
Boron	mg/l	0.5	0.45	10	0.75[b] 0.33[c]
Cadmium	μg/l	< 1.3	< 0.2	85	10
Chromium	μg/l	< 16	4	< 75	100
Cobalt	μg/l	< 3	3		50
Copper	μg/l	15	8	47	200
Fluoride	mg/l	1.3	1.25	4	1,000
Iron	μg/l	134	18	87	5,000
Lead	μg/l	< 5	< 3	40	
Manganese	μg/l	48	20	58	200
Molybdenum	μg/l	< 3	< 3		10
Nickel	μg/l	33	11	67	100
Selenium	μg/l	< 1.6	< 1	38	20

[a] According to EPA criteria.
[b] Recommended maximum concentration for irrigating citrus.
[c] Boron Class 1 for sensitive crops according to U.S. Department of Agriculture.

irrigation. The sodium concentration in the recovered water is similar to that of the recharge water (about 200 mg/l), so the cation exchange process was exhausted in the area surrounding the recharge basins. Adsorption of sodium and release of calcium and magnesium are still occurring in areas further away from the recharge basins. Potassium, which is found in relatively low concentrations in the recharge water, is still removed (30 percent) by SAT. Boron in the recharge water (0.5 mg/l) is only slightly removed (10 percent) by SAT. The concentrations of trace elements in the water after SAT are below the recommended maximum limits for irrigation water used continuously on all soils.

Conveyance of Recovered Water for Agricultural Reuse

Since November 1989, the recovered water from the Dan Region project has been transferred by a 100-km-long (62.5-mile-long) conveyance main called the Third Line to a distribution net for irrigated areas in the southern part of Israel. At the head of the conveyance system, the recovered water is disinfected by chlorination. Along the Third Line, there are four open operational reservoirs, each with a capacity of 13 to 26 million gallons. The reservoirs are used to compensate for changes in water pressure in the main pipe and to facilitate control of the system during peak demand periods. The retention time of water in the reservoirs is not constant and depends on operational restrictions. The monthly consumption of the recovered water conveyed by the Third Line during 1991 has ranged from 2.6 to 29 million gallons. During 2 years of operation, about 44,000 million gallons reclaimed water from the Dan Region project was reused for unrestricted irrigation in the southern part of Israel.

Summary

The soil-aquifer treatment (SAT) system as applied in the Dan Region Project is an efficient, low cost process ($0.03/m^3 for operation and maintenance only) for water reuse. Although the concentrations of several toxic substances in the recovered water are below the maximum permissible limits for drinking water, and turbidity is reduced by SAT, the recovered water is used only for nonpotable purposes. Overall, the recharge activities in the Dan Region illustrate how an unconventional source of water can be managed to increase the supply available for a variety of nonpotable uses. The very high quality of recovered water obtained after SAT makes the water suitable for agricultural uses (including unrestricted irrigation of vegetables to be eaten raw and livestock watering), industrial uses, nonpotable municipal uses (lawn irrigation and toilet flushing), and recreational uses. The main advantages of incorporating SAT are that it provides seasonal and multiyear underground storage, it is reliable, it provides a safety barrier, and it could lead to improved public acceptance of water reuse.

REFERENCES

Argo, D. G., and N. M. Cline. 1985. Groundwater Recharge Operations at Water Factory 21, Orange County, California. In Artificial Recharge of Groundwater, T. Asano ed. Boston, Mass.: Butterworth.

Aronson, D. A., and G. E. Seaburn. 1974. Appraisal of operating efficiency of recharge basins on Long Island, New York, in 1969. U.S. Geol. Surv. Water Supply Paper 2001-D, 22 pp.

Baird, R. B. 1987. GC-Negative ion CIMS and Ames mutagenicity assays of resins in advanced wastewater treatment facilities. In Advances in Sampling and Analysis of Organic Pollutants from water, I. H. Suffet and M. Malaiyandi, eds. Vol. 2, American Chemical Society, Advances in Chemistry, Washington, D.C.

Bouwer, E. J., P. L. McCarty, H. Bouwer, and R. C. Rice. 1984. Organic contaminant behavior during rapid infiltration of secondary wastewater at the Phoenix 23rd Avenue Project. Water Res. 18:463-472.

Bouwer, H. 1992. Agricultural and municipal use of wastewater. Water Sci. Technol. 26:1583-1591.

Bouwer, H., R. C. Rice, and E. D. Escarcega. 1974a. High-rate land treatment. I. Infiltration and hydraulic aspects of the Flushing Meadows Project. J. Water Pollut. Contr. Fed. 46:835-843.

Bouwer, H., J. C. Lance, and M. S. Riggs. 1974b. High-rate land treatment. II. Water quality and economic aspects of the Flushing Meadows Project. J. Water Pollut. Contr. Fed. 46:844-859.

Bouwer, H., R. C. Rice, J. C. Lance, and R. G. Gilbert. 1980. Rapid-infiltration research—The Flushing Meadows Project, Arizona. J. Water Pollut. Contr. Fed. 52:2457-2470.

Bouwer, H., and W. L. Chase, Jr. 1984. Water reuse in Phoenix, Arizona. Pp. 337-353 in Proceedings, Water Reuse Symposium III: Future of Water Reuse, San Diego, California, August 26-31. American Water Works Association.

Bouwer, H., and R. C. Rice. 1984. Renovation of wastewater at the 23rd Avenue rapid-infiltration project. J. Water Pollut. Contr. Fed. 56:76-83.

Bouwer, H. and Rice, R. C. 1989. Effect of water depth in groundwater recharge basins on infiltration. J. Irrig. and Drain. Engr., 115, 556-567.

Bradner, L. A. 1991. Water quality in the upper Floridan aquifer in the vicinity of drainage wells, Orlando, Florida. U.S. Geol. Surv. Water Resour. Invest. Rep. 90-4175, 57 pp.

Carlson, R. R., K. D. Lindstedt, E. R. Bennett, and R. B. Hartman. 1982. Rapid infiltration treatment of primary and secondary effluents. J. Water Pollut. Contr. Fed. 54:270-280.

Charbeneau, R. J. 1982. Groundwater resources of the Texas Rio Grande basin. Natur. Resour. J. 22(4):957-970.

German, E. R. 1989. Quantity and quality of stormwater runoff recharged to the Floridan aquifer system through two drainage wells in the Orlando, Florida, area. U.S. Geol. Surv. Water Supply Paper 2344, 51 pp.

Gilbert, R. G., C. P. Serba, R. C. Rice, H. Bouwer, C. Wallis, and J. L. Melnick. 1976. Virus and bacteria removal from wastewater by land treatment. Appl. Environ. Microbiol. 32:333-338.

Hartling, E. C. 1993. Impacts of the Montebello Forebay Groundwater Recharge Project. Bull. Calif. Water Pollut. Contr. Assoc. 29(3):14-26.

Kimrey, J. 0. 1978. Preliminary appraisal of the geohydrologic aspects of drainage wells, Orlando area, central Florida. U.S. Geol. Surv. Water Resour. Invest. Rep. 78-37, 24 pp.

Kimrey, J. 0., and L. D. Fayard. 1984. Geohydrologic reconnaissance of drainage wells in Florida. U.S. Geol. Surv. Water Resour. Invest. Rep. 84-4021, 67 pp.

Knorr, D. B. and T. Cliett. 1985. Proposed groundwater recharge at El Paso, Texas. In Artificial Recharge of Groundwater, T. Asano ed. Boston, Mass.: Butterworth.

Ku, H. F. H., and D. L. Simmons. 1986. Effect of urban stormwater runoff on ground water beneath recharge basins on Long Island, New York. U.S. Geol. Surv. Water Resour. Invest. Rep. 85-4088, 67 pp.

Lance, J. C., R. C. Rice, and R. G. Gilbert. 1980. Renovation of wastewater by soil columns flooded with primary effluent. J. Water Pollut. Contr. Fed. 52:381-388.

Leach, L. E., C. G. Enfield, and C. C. Harlin, Jr. 1980. Summary of Long-term Rapid Infiltration System Studies. EPA-600/2080-165. U.S. Environmental Protection Agency. Ada, Okla.

McCarty, P. L., M. Reinhard, N. L. Goodman, J. W. Graydon, G. D. Hopkins, K. E. Mortel-mans, and D. G. Argo. 1982. Advanced Treatment for Wastewater Reclamation at Water Factory 21. Techn. Paper No. 267. Department of Civil Engineering, Stanford University. Stanford, Calif.

McCarty, P. L., B. E. Rittman, and E. J. Bouwer. 1986. Microbiological processes affecting chemcial transformations in groundwater. Pp. 89-116, in Groundwater Pollution Microbiology, G. Bitton and C. P. Gerba, eds. New York: John Wiley.

Nellor, M. H., R. B. Baird, and J. R. Smyth. 1984. Health Effects Study—Final Report. NTIS No. PB-84191-568 County Sanitation Districts of Los Angeles County. Whittier, Calif.

Orange County Water District. 1991. Groundwater Management Plan. Orange County Water District. Fountain Valley, Calif.

Rice, R. C., and H. Bouwer. 1984. Soil-aquifer treatment using primary effluent. J. Water Pollut. Contr. Fed. 56:84-88.

Rice, R. C., and R. G. Gilbert. 1978. Land treatment of primary sewage effluent: Water and energy conservation. Pp. 33-36 in Hydrology and Water Resources in Arizona and the Southwest. Tucson, Ariz: University of Arizona Press.

Schiner, G. R., and E. R. German. 1983. Effects of recharge from drainage wells on quality of water in the Floridan aquifer in the Orlando area, central Florida. U.S. Geol. Surv. Water Resour. Invest. Rep. 82-4094, 124 pp.

Seaburn, G. E., and D. A. Aronson. 1974. Influence of recharge basins on the hydrology of Nassau and Suffolk counties, Long Island, New York. U.S. Geol. Surv. Water Supply Paper 2031, 66 pp.

Semmens, M. J., and T. K. Field. 1980. Coagulation: Experiences in organics removal. J. Am. Water Works Assoc. 72:476-483.

Sloss, E. M. 1993. Epidemiological assessment of groundwater recharge with reclaimed water in Los Angeles County. Proposal No. 93-019, submitted to the Water Replenishment District of Southern California by RAND, Santa Monica, Calif.

State of California. 1987. Report of the Scientific Advisory Panel on Groundwater Recharge with Reclaimed Wastewater. Prepared for the California State Water Resources Control Board, Department of Water Resources, and Department of Health Services. Sacramento, Calif.

Stringfield, V. T. 1933. Ground-water investigations in Florida. Flor. Geol. Surv. Geol. Bull. 11, 33 pp.

Thurman, E. M. 1979. Isolation, Characterization, and Geochemical Significance of Humic Substances from Groundwater. Ph.D. dissertation. University of Colorado, Boulder, Col.

Tibbals, C. H. 1990. Hydrology of the Floridan aquifer system in east central Florida.U.S. Geol. Surv. Prof. Paper 1403-E, 98 pp.

U.S. Environmental Protection Agency. 1992. Guidelines for Water Reuse. Technology Transfer Manual EPA/625/R-92/004. U.S. Environmental Protection Agency. Washington, D.C. 247 pp.

Wanielista, M. P., Y. A. Yousef, and S. J. Taylor. 1981. Stormwater Management to Improve Lake Water Quality. Submitted to Municipal Environmental Research Laboratory, Edison, N.J. Orlando, Fla.: University of Central Florida. 225 pp.

Wesner, G. M. 1991. Annual Report, Orange County Water District Wastewater Reclamation and Recharge Project, Calendar Year 1990. Prepared for Orange County Water District, Fountain Valley, Calif.

7

Conclusions and Recommendations

As demand for water increases, water managers and planners need to look widely for ways to improve water management and augment water supplies. The Committee on Ground Water Recharge concludes that artificial recharge can be one option in an integrated strategy to optimize total water resource management, and it believes that with pretreatment, soil-aquifer treatment, and posttreatment as appropriate for the source and site, impaired-quality water can be used as a source for artificial recharge of ground water aquifers.

Artificial recharge using source waters of impaired quality is a sound option where recharge in intended to control saltwater intrusion, reduce land subsidence, maintain stream baseflows, or similar in-ground functions. It is particularly well suited for nonpotable purposes, such as landscape irrigation, because health risks are minimal and public acceptance is high. Where the recharged water is to be used for potable purposes, the health risks and uncertainties are greater. In the past, the development of potable supplies has been guided by the principle that water supply should be taken from the most desirable source feasible, and the rationale for this dictate remains valid. Thus, although indirect potable reuse occurs throughout the nation and world wherever treated wastewater is discharged into a water course or underground and withdrawn downstream or downgradient for potable purposes, such sources are in general less desirable than using a higher quality source for potable purposes. However, when higher-quality, economically feasible sources are unavailable or insufficient, artificially recharged ground water may be an alternative for potable use.

The following conclusions and recommendations emerged from the committee's deliberations:

ARTIFICIAL RECHARGE: A VIABLE OPTION

Artificial recharge of ground water using source waters of impaired quality can be a viable way to augment regional water supplies—primarily for nonpotable purposes but for potable purposes under appropriate conditions—and at the same time provides an avenue for wastewater management.

Artificial recharge with waters of impaired quality has been practiced successfully in various parts of the United States and elsewhere for many years. Source water options include treated municipal wastewater, stormwater runoff, and irrigation return flow. Treated municipal wastewater and stormwater runoff are the two most commonly used sources; experience with the intentional use of irrigation return flow is scarce and not well documented. Recharge can be accomplished either through surface infiltration methods or through injection directly into the aquifer by wells. Hydrogeologic conditions, land availability, and the purpose of the recharge dictate the method of recharge, which in turn dictates the required pre-recharge treatment of the source water. Surface infiltration methods are used far more frequently than wells because of economic and operational considerations. However, well recharge is increasing because suitable sites for surface infiltration are not always available.

Ground water recovered from aquifers recharged with waters of impaired quality has been used for various purposes, ranging from landscape irrigation to potable supply. The desirability of using such waters for various purposes depends on the quality, availability, and cost of alternative sources of supply and varies considerably by site and source. One advantage of nonpotable reuse is that it releases other, higher quality sources for potable use.

A fundamental conclusion of this report is that impaired quality waters used to recharge ground water aquifers must receive a sufficiently high degree of pretreatment (prior to recharge) to minimize the extent of any degradation of ground water quality, as well as to minimize the need for any extensive post-treatment at the point of recovery. With surface infiltration systems, considerable quality improvements can be obtained as the water flows through the unsaturated zone to the aquifer; this soil-aquifer treatment (SAT) reduces pretreatment requirements.

Although some impacts of artificial recharge of ground water with source waters of impaired quality are not understood with complete certainty, experience with recharge projects has failed to show (within the limitations of toxicological testing) that water recovered from the aquifer poses greater health risks than currently acceptable potable water supplies. The state of our knowledge about artificial recharge using waters of impaired quality is more than sufficient to indicate that this technology offers particularly significant potential for all nonpotable uses. With proper pretreatment and posttreatment or dilution with native ground water, potable use also can be a viable option. These statements must, of course, be qualified by the fact that conditions vary from site to site, and thus the appropriateness of all recharge is site-specific. In particular, the quality of source waters, the hydrogeological setting, the costs of recharge facilities, and the availability and costs of alternative sources of supply will differ from situation to situation, and these factors must be considered when evaluating the feasibility of a specific artificial recharge project.

Recommendations

- Once artificial recharge has been deemed feasible as part of an integrated approach to regional water supply planning, the method of recharge chosen should be based on hydrogeologic conditions and the specific benefits sought from the recharge. In general, surface spreading offers the greatest engineering and operational advantages. Surface methods can accommodate waters of poorer quality and are simpler to design and operate than recharge wells, although certain conditions may require use of wells. Because surface spreading requires large amounts of land with permeable soil, it may not be feasible in densely populated areas or elsewhere where suitable land is expensive or unavailable. Injection wells require high quality source waters to avoid clogging problems and also because aquifers alone do not provide the same degree of treatment as soil-aquifer systems. Although there are indications of some water quality improvements within aquifers, considerable pretreatment is necessary if the source water to be used in wells is of impaired quality.
- Artificial recharge using water of impaired quality offers particularly significant potential for nonpotable uses. Nonpotable reuse can help reduce demand on limited fresh water sources at minimal health risk; it is widely practiced and achieves good public acceptance. Potable reuse is equally possible to engineer, but the health risks may be greater and public acceptance is less certain. In either approach, but especially where potable reuse is considered, careful preproject study and planning is required.

POTENTIAL IMPAIRED QUALITY SOURCES

Three main types of impaired quality waters are potentially available for ground water recharge—treated municipal wastewater, stormwater runoff, and irrigation return flow. Of these, treated municipal wastewater is usually the most consistent in terms of quality and availability. Stormwater runoff from residential areas generally is of acceptable quality for most recharge operations, but at some times and places it may be heavily con-

taminated, and its availability is variable and unpredictable. Irrigation return flow exhibits wide variations in quality and is sometimes seriously contaminated, and thus usually is not a desirable source of water for recharge.

Treated municipal wastewater is by far the most consistent impaired-quality water source, both spatially and temporally and in terms of quality and quantity. One exception to this generalization is municipal wastewater and stormwater commingled in a combined wastewater collection system, and another occurs when industrial wastewater is discharged to the municipal wastewater collection system. The quality of treated municipal wastewater has been characterized for various levels of treatment to meet regulations pertaining to the disposal of sewage effluent and to allow use of the effluent for recharge and other purposes. The characterization of the quality of stormwater runoff and irrigation return flow is far less comprehensive because general assessments of stormwater and irrigation flow quality must be drawn from a much less systematic and comprehensive database than is available for treated municipal wastewater.

Attempts to use impaired-quality water sources for artificial recharge should be conservative. For this reason, the choice of source water and the degree of treatment necessary for the intended use are critical. Although soil-aquifer treatment improves water quality, the precise level of soil-aquifer treatment achieved often is unpredictable and very difficult to monitor, suggesting that the most reasonable course is to require the best possible source water and use impaired-quality sources only in appropriate circumstances. Municipal wastewater that has undergone at least secondary treatment provides a widely available source water that contains levels of many contaminants within the treatment and removal capability of well-designed and well-managed soil-aquifer treatment systems.

Recommendations

- Based on current information, municipal wastewater used for artificial recharge should receive at least secondary treatment. Municipal wastewater that has received only primary treatment may be adequate for the recharge of nonpotable ground water in certain areas, but use of primary effluent should not be considered without implementation of a site-specific demonstration study.
- Certain impaired-quality waters, such as irrigation return flow, industrial wastewater, and stormwater runoff from industrial areas, generally should not be regarded as suitable sources for artificial recharge. Exceptions might be identified, but only after careful characterization of source water quality on a case-by-case basis. Other types of stormwater runoff to avoid include: most dry weather storm drainage flows, salt-laden snowmelt flows, and flows originating from certain commercial facilities, such as vehicle service areas. Construction site

runoff also should be avoided to prevent clogging of recharge facilities with eroded soil and other debris.

HUMAN HEALTH CONCERNS

The principal concern with regard to artificial recharge using waters of impaired quality for potable purposes is the protection of human health. Several major studies employing state-of-the-art methods for organic analysis and toxicological testing show that well-managed recharge projects produce recovered water of essentially the same quality from a health perspective as water from other acceptable sources. However, there are uncertainties in identifying potentially toxic constituents and pathogenic agents in the methodologies used in these studies, and thus potable reuse should only be considered when better quality sources are unavailable.

Recharge projects in the United States and elsewhere have provided analytical data on the chemicals and microorganisms found in treated municipal wastewater before and after recharge and soil-aquifer treatment. The concentrations of these constituents are highly variable and are dependent on the source of water and the specific sites involved. Although the database is not large, the available information does provide a basis for a limited assessment of the potential adverse health impacts when the recovered water is used as a potable supply.

All methodologies have inherent limitations, but on the basis of available information there is no indication that the health risks from water recovered after recharge of treated municipal wastewater are greater than those from existing water supplies, or that the concentrations of chemicals or microorganisms are higher than those established in drinking water standards by the Environmental Protection Agency (EPA). Uncertainties exist, however, where data are not available.

Drinking water from ground water recharge is regulated by EPA in the same fashion as drinking water from other ground water sources. Comparison with existing drinking water standards is one common and convenient approach for evaluating the quality of the recovered water. Other criteria to estimate the health risks from extracted ground water, such as health advisories developed by EPA, are useful. Artificially recharged ground water used for potable supplies need not be subject to stricter water quality requirements than conventional water supplies; however, as stated elsewhere in this report, water quality monitoring and operations management should be more stringent for recharge systems intended for potable reuse.

Assessing the risk to an individual from pathogens in ground water that has been recharged with impaired-quality water is difficult. Bacteria and parasites generally are removed to a greater extent than are enteric viruses during infiltration through soils; thus viruses may be of greater concern when there is human exposure to recovered water.

Disinfection by-products (DBPs) are of concern in artificially recharged ground water systems used for potable water, as they are in water supplies drawn from surface or naturally recharged ground waters. The nature and toxicity of such DBPs have been most widely studied for chlorine disinfection of potable water supplies. However, the possible differences in the nature and quantities of DBPs resulting from the disinfection of impaired-quality waters that may be used for ground water recharge have not been studied thoroughly.

A key issue in developing any potable water supply, including ground water recharge systems, is the need to balance the risks in using chemical disinfectants to reduce the number of pathogenic microorganisms with those associated with the DBPs formed in the process. As a crude comparison, it has been estimated that the probability of mortality from pathogenic microorganisms in improperly disinfected drinking water would exceed the carcinogenic risks introduced by chlorine as much as 1000-fold. Chlorination has been the most widely used disinfectant of highly treated municipal wastewater for ground water recharge and other uses, but other disinfection processes, including the use of ultraviolet radiation, are increasingly being assessed. Although the mix of DBPs formed from the use of these other disinfection processes requires more study, these alternatives may be more efficacious than chlorination in minimizing the health risks from ground water recharge due to pathogenic microorganisms and DBPs.

Although the health risks associated with potable use of recovered water are likely to be minimal and may be mitigated by sound design and operation of treatment facilities, this committee believes that the best available water sources should be used for potable purposes whenever possible in preference to ground water recharged with impaired-quality source water. Under conditions of increasing water scarcity, however, economic and practical considerations may dictate the use of lesser-quality source waters in recharge of ground water that ultimately serves potable purposes.

Recommendations

- Disinfection of treated municipal wastewater prior to recharge should be managed so as to minimize the formation of disinfection by-products. Alternatives to chlorination include disinfection with ultraviolet radiation and the use of other chemical disinfectants. However, additional research should be undertaken on pathogen removal and formation of disinfection by-products before alternative disinfectants can be classified as conclusively superior to chlorine.

- Recovered water must be monitored carefully to provide assurance that pathogenic microorganisms and toxic chemicals do not occur at concentrations that might exceed drinking water standards or other water quality parameters established specifically for reclaimed water which consider the nature of the source water. The outcomes of existing studies of potable use of recovered ground water recharged with treated municipal wastewater suggest that additional epidemiological, in-vivo, or short-term toxicological studies would be of marginal value. As long as the recovered water meets drinking water standards and other water quality limits specified for the site, and there is no evidence from monitoring of constituents that pose undesirable health risks, additional toxicological testing is unnecessary. If the extracted water is uncertain for any reason, it should not be considered for potable reuse.

- There are significant uncertainties associated with the transport and fate of viruses in recharged aquifers. These uncertainties make it difficult to determine the levels of risk of any infectious agents still contained in the disinfected wastewater. Thus, additional research should be undertaken on the transport and fate of viruses in recharged aquifers to allow improved assessments of the possible health risks and needs for post-extraction disinfection associated with such systems.

- Artificial recharge of ground water with waters of impaired quality should be used to augment water supplies for potable uses only when better-quality sources are not available, subject to thorough consideration of health effects and depending on economic and practical considerations.

SYSTEM MANAGEMENT AND MONITORING

Protecting public health and the sustainability of soil-aquifer systems will require careful planning, operation, and management of recharge systems. Under appropriate conditions, the soil-aquifer system has the capacity to remove certain chemicals and pathogens and can therefore be an effective component in ground water recharge and water reuse systems. However, the processes through which removal occurs are not completely efficient in natural settings, and not all constituents are retained or degraded to the same extent. In addition, strategies that may enhance the removal of one chemical or pathogen can decrease the efficiency of removal of another.

The protection of both human health and the environment are goals in any recharge system, and both require careful attention to system management and monitoring. The use of recharge technologies may have impacts on the environment, and the presence of these impacts as well as their magnitude will vary from situation to situation. The careful operation of wastewater treatment systems for ground water recharge has shown that the use of various processes can reduce the concentrations of nitrogen, phosphorus, heavy metals, organic chemicals, suspended solids, and pathogenic microorganisms in the effluent used for recharge. However, even when treated to a high degree, effluent disinfected with high chlorine doses may contain disinfection by-products (DBPs). If the water is to be used for drinking, it is critical that the formation of DBPs be minimized by focusing on the nature and location of the disinfection process and by ensuring the optimal combination of pre- and post-disinfection.

The long-term viability of any soil-aquifer treatment system will depend on the specific nature of the source water and its treatment, the soil, and the receiving aquifer. Some challenges to sustainability, such as clogging caused by suspended material and biological activity, can be managed. Others, such as the attenuation of viruses and other pathogens and the accumulation of metals, phosphorus, organic compounds, and other constituents, may be more problematical. Thus, monitoring of the recharge system is needed to evaluate its long-term behavior and to formulate appropriate actions when needed. Of the two methods of artificial recharge—surface infiltration and well recharge—the latter requires source water of much higher quality.

Artificial recharge is an established technology, and while there is always room for research and improvement in areas such as how to optimize the process, minimize costs, and maximize safety, the greatest remaining uncertainties relate to the potential implications for human health. In particular, research into the fate and transport of chemicals in recharge waters, removal mechanisms for organic constituents in ground water, and use of alternative disinfection techniques offer potential. There is also a need for efforts to synthesize existing performance data.

Recommendations

- Assessments of the feasibility of any recharge technology should include analyses of the possible impacts of the use of the system on the environment.
- Monitoring of recharge water should be undertaken as it moves toward points of recovery. This is critical to help ensure that water quality is maintained, to provide early warning of unexpected problems, and to help maintain the long-term viability of the treatment system.

ECONOMIC CONSIDERATIONS

Artificial recharge opportunities need to be evaluated within the overall context of available water supplies, existing and projected water demands, and related costs and benefits to ensure that the opportunity is economically justified.

Artificial recharge is but one option for augmenting water supplies for potable and nonpotable uses. Most communities and regions will have several such options, and it will be important to evaluate all options on the same basis if the least costly option is to be identified. The cost of otherwise disposing of wastewaters is one factor to be taken into account. The benefits associated with the ultimate uses to which recharge water is put should be equal to or greater than this cost. The temptation to subsidize recharge operations in order to ensure that the water will be used or to make its price more attractive should be resisted. Subsidies result in distorted water allocation and inefficiencies in use.

Recommendation

- The price of recovered water should reflect the true cost of making the water available to ensure that the water is used efficiently. The costs of recharge operations should not be subsidized to make this water source more attractive than it would otherwise be.

LEGAL AND INSTITUTIONAL CONSIDERATION

The development of institutional arrangements governing artificial recharge is critical in determining the extent to which water supplies will ultimately be available from recharge with waters of impaired quality. The institutions need to be capable of formulating policies to protect public health and environmental amenities while not imposing inappropriate or inefficient controls on this potentially important form of water resource management. Federal leadership is needed if the full promise of artificial ground water recharge is to be realized.

The feasibility of artificial recharge could be improved by appropriate institutional arrangements to govern and regulate the recharge of impaired quality water and the uses of recovered water. If the regulatory process is itself uncertain, there will be general underinvestment in recharge facilities. Underinvestment will also result from processes that are too conservative. Processes that fail to account adequately for public health and welfare concerns and impacts on the environment are also likely to result in the imposition of unnecessary social costs.

Artificial recharge of ground water will clearly be impeded where title to the recovered water is unclear or undefined. Similarly, clear legal rights to source waters will be required if recharge is to occur. Although institutional arrangements may need to be flexible and may vary from situation to situation, a significant first step to establishing an institutional environment that will foster recharge will be to ensure that these rights are clearly defined. This task will virtually always fall to the states.

Interests in environmental amenities and the benefits they provide are widely distributed and diffuse. Often, no one individual or group has enough stake in the preservation of environmental amenities to permit or motivate it to intervene in the regulatory process. The National Environmental Policy Act applies only to federal activities, and not all states have comparable statutes. Inasmuch as artificial recharge operations can have significant environmental impacts, it is important that regulatory processes be designed so as to ensure that there is adequate assessment, evaluation, and review of the environmental impacts of artificial recharge projects.

In the absence of federal leadership, states will be compelled to develop standards and regulatory schemes independently. Although the needs, economies, and opportunities of states vary widely and thus state-level leadership and flexibility of approach is important, many states lack the resources to develop adequate regulation and guidance. Moreover, the development of 50 independent sets of standards is inefficient and is unlikely to result in optimal development of artificial recharge operations nationwide. The federal government has the capability to provide technical assistance to the states and help develop model statutes and guidelines to assist the states in developing their own policies. As more experience is gained with the regulatory aspects of recharge, the partition of responsibilities between the national and state governments should be reexamined.

Recommendations

- As a first step in developing institutional arrangements that will foster artificial recharge as a means of augmenting water supplies, states should move to clarify legal rights to source waters and recovered waters for artificial recharge operations.
- In addition to ensuring the protection of public health related to the consumption of recovered water, when developing regulatory policies states should make explicit provision for the evaluation of project sustainability and environmental impacts of artificial recharge projects.
- Regulatory processes should ensure that environmental impacts and other third party effects are adequately accounted for in the design and operation of artificial recharge projects.
- The federal government should assume leadership in supporting the development of artificial recharge with municipal wastewater and other suitable impaired-quality water sources by providing technical assistance to the states and by developing model statutes and guidelines.

Appendixes

APPPENDIX
A

Acknowledgements

The generation of a report such as this one takes input from many people. The committee wishes to extend its sincere appreciation to all the people who shared their time and experience with us during the preparation of this report. This includes a variety of people who joined us at meetings, hosted our site visits, led us on field trips, and contributed their knowledge to our efforts. In particular, we would like to thank the following people for their contributions:

Takashi Asano, consultant, Davis, California
Richard Atwater, Central Basin Municipal Water District Carson, California
Robert Bastian, EPA, Office of Municipal Pollution Control, Washington, D.C.
Richard Bull, Washington State University, Pullman
Shirley Clark, University of Alabama, Birmingham
Robert C. Cooper, University of California, Berkeley
Joseph Cotruvo, EPA Office of Pollution Prevention and Toxics, Washington D.C.
Carl Enfield, EPA, Robert S. Kerr Environmental Research Laboratory, Ada, Oklahoma
Richard Engelbrecht, University of Illinois at Urbana-Champaign
Charles Gerba, University of Arizona, Tucson
Bruce Glenn, Bureau of Reclamation, Denver, Colorado
Jeff Helsley, Water Replenishment District of Southern California, Cerritos
Jim Hunt, University of California, Berkeley
Helen Ingram, University of Arizona, Tucson
Bruce Johnson, Tucson Water, Tucson, Arizona

Charles C. Johnson, consultant, Bethesda, Maryland
Ivan Johnson, consultant, Arvada, Colorado
A. Kanarek, Mekorot Water Company, Israel
Marie Light, Tucson Water, Tucson, Arizona
M. Michail, Mekorot Water Company, Israel
Dan Okun, University of North Carolina, Chapel Hill
Ken Reich, West Basin Municipal Water District, Carson, California
Norbert S. Ries, Bureau of Reclamation, Denver, Colorado
Martin Rigby, Orange County Water District, Fountain Valley, California
Paul Roberts, Stanford University, Stanford, California
Rusty Schuster, Bureau of Reclamation, Denver, Colorado
Victoria J. Tschinkel, Landers and Parsons, Tallahassee, Florida
Nancy Ward, Tucson Water, Tucson, Arizona
Gary Westerhoff, Malcolm Pirnie, Inc., White Plains, New York
L.G. Wilson, University of Arizona, Tucson

APPENDIX
B
Biographical Sketches of Committee Members

Julian Andelman, the committee's chair, is professor of Environmental Health in the Department of Environmental and Occupational Health, as well as professor of Civil Engineering at the School of Engineering at University of Pittsburgh. He received his A.B. in biochemical sciences from Harvard College and his Ph.D. in physical chemistry from Polytechnic Institute of Brooklyn. His research focuses on the behavior of chemicals in treated and natural waters, and the transport of volatile chemicals from water to air in indoor systems, as well as the human exposures that result.

Herman Bouwer is chief engineer with the U.S. Water Conservation Laboratory of the U.S. Department of Agriculture. Dr. Bouwer received B.S. and M.S. degrees in land drainage and irrigation from the National Agricultural University of Wageningen, the Netherlands, and his Ph.D. in hydrology and agricultural water management from Cornell University. He is adjunct professor at Arizona State University and the University of Arizona. His research focuses on in situ measurement of hydraulic properties of soils, vadose zones, and aquifers; characterization of underground flow systems; renovation of sewage effluent by ground water recharge; effect of irrigated agriculture on ground water; and effect of ground water pumping on stream flow.

Randall Charbeneau is director at the Center for Research in Water Resources at the University of Texas-Austin, as well as Professor in the University's Department of Civil Engineering. He holds B.S., M.S., and Ph.D. degrees in engineering from the University of Michigan, Oregon State University, and Stanford University, respectively. Dr. Charbeneau has served on several ground water research advisory panels, including one on ground water contamination for the Geophysical Research Board of the National Research Council. His research interests include hazardous waste disposal and ground water transport modeling.

Russell Christman is professor of environmental sciences at the University of North Carolina-Chapel Hill. Dr. Christman received his B.S., M.S., and Ph.D. in chemistry from the University of Florida. He has taught sanitary and civil engineering at the University of Washington, where he also served as Associate Professor of the Applied Sciences. His research focuses on the chemical structures of natural product organic materials in water; methods of organic analysis in water samples; and mechanisms of colloidal destabilization with hydrolysis products of aluminum III.

James Crook is director of water reuse for the firm Black & Veatch. He was previously with Camp Dresser & McKee Inc. Prior to that he was with the California Department of Health Services, where he directed the department's water reclamation and reuse program. Dr. Crook has served on several water reuse advisory panels and has been an advisor to the National Sanitation Foundation, Pan American Health Organization, United Nations Development Programme, and U.S. Agency for International Development. He was the principal author of water reuse guidelines published by the U.S. Environmental Protection Agency and has assisted in the development of water reclamation and reuse criteria for several states. Dr. Crook received his B.S. in civil engineering from the University of Massachusetts; he received his M.S. and Ph.D. in environmental engineering from the University of Cincinnati.

Anna Fan is chief, Pesticide and Environmental Toxicology Section, Office of Environmental Health and Hazard Assessment, California Environmental Protection Agency, State of California. She received her Ph.D. in toxicology from Utah State University. Dr. Fan has worked to develop water standards for the state. Her responsibilities include evaluating and establishing California's safe drinking water standards, conducting risk-based assessments of chemical contamination in various environmental media, and conducting epidemiological investigations of pesticide health effects. Dr. Fan chairs a public health working group and has worked on contaminants in drainage water in the San Joaquin Valley. She is a member of the Drinking Water Committee, Science Advisory Board, United States Environmental Protection Agency. She is also an adjunct professor at San Jose State University.

Denise Fort is director of the Water Resources Administration Program at the University of New Mexico and a member of the faculty of the School of Law. Previously she was a consultant with the Natural Heritage Institute in California. She has served as attorney for the New Mexico Public Interest Group, the Southwest Research and Information Center, and the Taxation and Revenue Department of New Mexico. She was also Cabinet Secretary of the New Mexico Department of Finance and Administration where she was responsible for management of the state's budget, fiscal controls, public school and local government finance, capital bonding, and planning. She later became director of the Environmental Improvement Division of New Mexico, where she was responsible for the administration of the State's environmental laws. She received her B.S. from St. John's College in Santa Fe, New Mexico and her J.D. from the Catholic University of America.

Wilford Gardner has recently retired as Dean, College of Natural Resources, University of California at Berkeley. He received a Ph.D. in physics from Iowa State College in 1953. Dr. Gardner was previously with the Department of Soils, Water, and Engineering at the University of Arizona, Tucson. He has been a National Science Foundation senior fellow at Cambridge University and a Fulbright Lecturer, University of Ghent. Dr. Gardner is a member of the National Academy of Sciences and the Water Science and Technology Board. His research has been in measurement of soil moisture by neutron scattering; soil physics; movement of fluids in porous media; soil-water plant relations; soil salinity; plant biophysics; and environmental physics.

William Jury is professor of soil physics and chair of the Department of Soil and Environmental Sciences at the University of California-Riverside. He received his Ph.D. in physics from the University of Wisconsin in 1973. Dr. Jury's principal research interests are measurement and modeling of organic and inorganic chemical movement and reactions in field soils; development and testing of organic chemical screening models; spatial variability of soil physical and chemical properties; and assessing volatilization losses of organic compounds.

David Miller is president, CEO, and chairman of the board for the firm Geraghty and Miller, Inc. He received a B.A. and M.S. in geology from Colby College and Columbia University in 1953. He is well know as an expert in applied ground water hydrology, and his career includes service with the USGS and consulting firms. Mr. Miller was a founding member of the WSTB. He has worked on more than 200 projects for industries and utilities. He is now working on treating contaminated ground water at Superfund sites.

Robert Pitt is associate professor in the Civil and Environmental Engineering Department and the Environmental Health Science Department at the University of Alabama at Birmingham. He received his B.S. in engineering science from Humboldt State University in 1970, his M.S.C.E. in environmental engineering/hydraulic engineering from San Jose State University in 1971, and his Ph.D. in civil and environmental engineering from the University of Wisconsin-Madison in 1987. His principal research interests include investigating urban runoff pollutant effects, sources and controls.

Henry Vaux, Jr. is Associate Vice President-Agricultural and Natural Resource Programs for the University of California systemwide. He previously served as Director of the University of California Water Resources Center and as a professor of resource economics at the University of California, Riverside. His principal research interests are the economics of water use and water quality. Prior to joining the University he worked at the Office of Management and

Budget and was on the staff of the National Water Commission. He received a Ph.D. in natural resource economics from the University of Michigan in 1973.

John Vecchioli is a hydrologist with the U.S. Geological Survey's Water Resources Division in Tallahassee, Florida, and currently serves as Chief of the Florida District Program. Previously, he was responsible for quality assurance of all technical aspects of ground water programs in Florida. He was project chief of a study of hydraulic and geochemical aspects of waste injection in Florida. Also, while with the Survey on Long Island, New York, he was project chief for a study involved in planning and supervision of a multidisciplinary effort to evaluate the hydraulic and geochemical impact of artificial recharge of treated wastewater through injection wells. Mr. Vecchioli received his B.S. and M.S. in geology from Rutgers University in 1956 and 1957, respectively. He is a member of the American Geophysical Union, Geological Society of America, National Ground Water Association, American Institute of Professional Geologists, International Association of Hydrogeologists, and Sigma Xi.

Marylynn Yates is associate professor of environmental microbiology with the Department of Soil and Environmental Sciences, University of California-Riverside. Dr. Yates received her B.S. in nursing from the University of Wisconsin-Madison, her M.S. in chemistry from the New Mexico Institute of Mining and Technology, and her Ph.D. in microbiology and immunology from the University of Arizona. She is a member of the American Waterworks Association technical working group looking at EPA's disinfection rule. Her research interests include the fate of microorganisms in soil, surface water, ground water, and wastewater treatment; modeling the fate and transport of microorganisms in the subsurface; ground water contamination; and waterborne disease outbreaks.

APPENDIX
C

Glossary

Acre-foot — The volume of water required to cover 1 acre of land to a depth of 1 foot. Equal to 1.23 megaliters (ML) or 1,230 cubic meters (m^3).

Advanced wastewater treatment — Any physical, chemical or biological treatment process used to accomplish a degree of treatment greater than that achieved by secondary treatment.

Aquifer — A formation, group of formations, or part of a formation that contains sufficient saturated permeable material to yield significant quantities of water to wells and springs.

Basin — (1) *Hydrology:* The area drained by a river and its tributaries. (2) *Irrigation:* A level plot or field, surrounded by dikes, which may be flood irrigated. (3) *Runoff control:* A catchment constructed to contain and slow runoff to permit the settling and collection of soil material transported by overland and rill runoff flows.

Benefit-cost ratio — An economic indicator of the efficiency of a proposed project, computed by dividing benefits by costs; usually, both the benefits and the costs are discounted so that the ratio reflects efficiency in terms of the present value of future benefits and costs.

CFU — Colony-forming unit.
COD — Chemical oxygen demand.
Contaminant — An undesirable substance not normally present or an unusually high concentration of a naturally occurring substance in water or soil.

Detection limit — A number of different detection limits have been defined: IDL (instrument detection limit), is the constituent concentration that produces a signal greater than five times the signal to noise ratio of the instrument; MDL (method detection limit) is the constituent concentration that, when processed through a complete method, produces a signal with a 99 percent probability that it is different from a blank; PQL (practical quantification limit) is the lowest constituent concentration achievable among laboratories within specified limits during routine laboratory operations. The ratios of these limits are approximately: IDL:MDL:PQL = 1:4:20.

Disinfection by-products — A range of organic and inorganic products resulting from the reaction of disinfecting oxidants with natural aquatic organic material reductants in water systems. The number and nature of all products are not precisely known at present, and vary with type of disinfectant employed. Some of the chlorination by-products are mutagenic (3-chloro-4-(dichloromethyl)-5-hydroxy-2(5H)-furanone) in the histidine reversion assay, and some are suspected animal carcinogens (Chloroform, and perhaps the entire family of trihalomethanes). Many disinfection by-products are of little or no toxicological significance, such as the non-halogen containing organic oxidation products and the simple halide ion reduction products.

DOC — Dissolved organic carbon.

Domestic wastewater — Sewage derived principally from human sources.

Enteric viruses — A large group of viruses that are characterized by the fact that they replicate in the intestinal tract and are therefore present in fecal material.

Enteroviruses — A specific group of enteric viruses that includes polioviruses, echoviruses, coxsackie viruses.

Entries to storm drainage — Water (relatively clean or polluted) discharged into a stormwater drain from sources such as, but not limited to, direct industrial or sanitary wastewater connections, roof leaders, yard and area drains, cooling water connections, manhole covers, ground water or subterraneous stormwater infiltration, etc.

Flow line — The general path that a particle of water follows under laminar flow conditions.

Fomites — inanimate objects that might be contaminated with infectious organisms and thus serve to transmit disease.

Ground water — That part of the subsurface water that is in the saturated zone.

Ground water mining — The withdrawal of ground water through wells, resulting in a lowering of the ground water table at a rate faster than the rate at which the ground water reservoir can be recharged.

APPENDIX C 281

Hydraulic conductivity — A proportionality constat relating hydraulic gradient to specific discharge which for an isotropic medium and homogeneous fluid, equals the volume of water at the existing kinematic viscosity that will move in unit time under a unit hydraulic gradient through a unit area measured at right angles to the direction of flow.

Head — The pressure of a fluid on a given area, at a given point caused by the height of the fluid surface above the point. Also, water-level elevation in a well, or elevation to which the water of a flowing artesian well will rise in a pipe extended high enough to stop the flow.

Hydraulic loading rate — In recharge, the average infiltration into a recharge basin expressed over time, including flooded, dry, and cleaning cycles for the basin.

Infiltration rate — Generally, the rate at which a soil under specified conditions can absorb falling rain or melting snow; in recharge, the rate at which water drains into the ground when a recharge basin is flooded, expressed in depth of water per unit time.

Injection well — Well used for emplacing fluids into the subsurface.

Irrigation return flow — The part of applied water that is not consumed by evapotrans-piration and that migrates to an aquifer or surface water body.

NTU — Nephelometric turbidity units.

Pathogen — A disease-causing microorganism.

Permeability — The property or capacity of a porous rock, sediment, or soil for transmitting a fluid without impairment of the structure of the medium; a measure of the relative ease of flow under unequal pressure.

PFU — Plaque-forming unit.

Porosity — The ratio of void volume to total volume of a porous medium.

Potable reuse, direct — Occurs when there is a piped connection of water reclaimed from wastewater to a potable water supply distribution system or a water treatment plant.

Potable reuse, planned indirect — Occurs when wastewater effluent is discharged to a water source with the intent of subsequently reusing the water rather than as a means of disposal.

Potable reuse, unplanned indirect — Occurs when a water supply is withdrawn for potable purposes from a natural surface or underground water source that is fed in part by the discharge of a wastewater effluent. The wastewater effluent is discharged to the water source as a means of disposal and subsequent reuse of the effluent is a byproduct of the disposal plan.

Potable water — Water that has been treated to be or is naturally suitable for drinking.

Pretreatment — As used in this report, any treatment (e.g., the removal of material such as gross solids, grit, grease, metals, toxicants, etc. or treatment such as aeration, pH adjustment, etc.) to improve the quality of a wastewater prior to recharge. This can also refer to the initial treatment processes of a sewage treatment plant.

Recharge area (ground water) — An area in which water infiltrates the ground and reaches the zone of saturation.

Recharge capacity — The ability of the soils and underlying material to allow precipitation and runoff to infiltrate and reach the zone of saturation.

Recharge, incidental — Water that infiltrates to the water table by seepage from manmade works such as water supply lines, sewage systems, septic tank leach fields, etc. which are unintended sources of ground water recharge.

Recharge, natural — The replenishment of ground water by downward infiltration of water from rainfall, streams, and other natural sources of water.

Recharge, planned (also called artificial) — As used in this report refers to any active and artificial means of enhancing natural recharge. It includes spreading basins, recharge pits, injection wells, and other direct means of recharging ground water basins. The purposeful reinjection of low quality water for disposal purposes, such as related to oil and gas recovery, is not intended for future beneficial use and is not included within this definition.

Reclaimed water — Wastewater made fit for reuse for potable or nonpotable purposes.

SAT — Soil-aquifer treatment refers to processes that occur in the soil and aquifer that act to remove or reduce chemical and biological constituents of concern.

Saturated zone — That part of the earth's crust beneath the regional water table in which all voids, large and small, are filled with water under pressure greater than atmospheric.

Secondary porosity — The porosity developed in a rock formation after its deposition or emplacement, either through natural processes of dissolution or stress distortion, or artificially through acidization or the mechanical injection of coarse sand.

Sheet flow — Overland flow in a relatively thin sheet of generally uniform thickness.

Site characterization — A general term applied to the investigation activities at a specific location that examines natural phenomena and human-induced conditions important to the resolution of environmental, safety and water-resource issues.

Stormwater runoff — Water resulting from precipitation which either infiltrates into the ground, impounds/puddles, or runs freely from the surface, or is captured by storm drainage, a combined sewer, and to a limited degree, by sanitary sewer facilities.

Tertiary treatment — The treatment of wastewater beyond the secondary or biological. Terms normally implies the removal of nutrients, such as phosphorus and nitrogen, and of a high percentage of suspended solids. Term now being replaced by preferable term, advanced waste treatment.
TDS — Total dissolved solids.
TOC — Total organic carbon.

Unsaturated zone — The zone between the land surface and the regional water table. Generally, water in this zone is under less than atmospheric pressure, and some of the voids may contain air or other gases at atmospheric pressure. Beneath flooded areas or in perched water bodies the water pressure locally may be greater than atmospheric.

Vadose zone — See *unsaturated zone*.

Wastewater — Water that carries wastes from homes, businesses, and industries; a mixture of water and dissolved or suspended solids.
Water quality — The chemical, physical, and biological condition of water related to a beneficial use.
Water resource — The supply of ground and surface water in a given area.
Watershed — A geographic region (area of land) within which precipitation drains into a particular river, drainage system or body of water that has one specific delivery point.